δ Delta

Single and Multiple-Digit Division

Instruction Manual

by Steven P. Demme

1-888-854-MATH (6284) - mathusee.com
sales@mathusee.com

Delta Instruction Manual: Single and Multiple-Digit Division
©2012 Math-U-See, Inc.
Published and distributed by Demme Learning

mathusee.com

1-888-854-6284 or +1 717-283-1448 | demmelearning.com
Lancaster, Pennsylvania USA

ISBN 978-1-60826-082-9
Revision Code 1219-B

Printed in the United States of America by CJK Group
 2 3 4 5 6 7 8 9 10

For information regarding CPSIA on this printed material call: 1-888-854-6284
and provide reference #1219-12102020

Building Understanding in Teachers and Students to Nurture a Lifelong Love of Learning

At Math-U-See, our goal is to build understanding for all students.

We believe that education should be relevant, skills-based, and built on previous learning. Because students have a variety of learning styles, we believe education should be multi-sensory. While some memorization is necessary to learn math facts and formulas, students also must be able to apply this knowledge in real-life situations.

Math-U-See is proud to partner with teachers and parents as we use these principles of education to **build lifelong learners.**

Curriculum Sequence

$\int$	**Calculus**
cos	**PreCalculus** with Trigonometry
xy	**Algebra 2**
Δ	**Geometry**
x^2	**Algebra 1**
x	**Pre-Algebra**
ζ	**Zeta** Decimals and Percents
ε	**Epsilon** Fractions
δ	**Delta** Division
γ	**Gamma** Multiplication
β	**Beta** Multiple-Digit Addition and Subtraction
α	**Alpha** Single-Digit Addition and Subtraction
P	**Primer** Introducing Math

Math-U-See is a complete, K-12 math curriculum that uses manipulatives to illustrate and teach math concepts. We strive toward "Building Understanding" by using a mastery-based approach suitable for all levels and learning preferences. While each book concentrates on a specific theme, other math topics are introduced where appropriate. Subsequent books continuously review and integrate topics and concepts presented in previous levels.

Where to Start
Because Math-U-See is mastery-based, students may start at any level. We use the Greek alphabet to show the sequence of concepts taught rather than the grade level. Go to mathusee.com for more placement help.

Each level builds on previously learned skills to prepare a solid foundation so the student is then ready to apply these concepts to algebra and other upper-level courses.

Major concepts and skills for Delta:
- Using strategies based on place value and properties of operations to divide
- Understanding division as solving for an unknown factor
- Fluently dividing any combination of whole numbers
- Solving abstract and real-world problems involving all four operations
- Interpreting remainders in short and long division
- Understanding fraction notation in light of division

Additional concepts and skills for Delta:
- Reading and writing Roman numerals
- Dividing, multiplying, adding, and subtracting U.S. currency and standard units of measure
- Understanding angle measure and geometric shapes including points, segments, rays, and lines
- Classifying shapes based on defining attributes
- Understanding and computing area and volume

Find more information and products at mathusee.com

Contents

HOW TO USE

Five Minutes for Success

Welcome to Delta. I believe you will have a positive experience with the unique Math-U-See approach to teaching math. These first few pages explain the essence of this methodology, which has worked for thousands of students and teachers. I hope you will take five minutes and read through these steps carefully.

The student should have a thorough grasp of addition, subtraction, and multiplication.

If you are using the program properly and still need additional help, you may visit us online at mathusee.com or call us at 888-854-6284. **–Steve Demme**

The Goal of Math-U-See

The underlying assumption or premise of Math-U-See is that the reason we study math is to apply math in everyday situations. Our goal is to help produce confident problem solvers who enjoy the study of math. These are students who learn their math facts, rules, and formulas and are able to use this knowledge to solve word problems and real-life applications. Therefore, the study of math is much more than simply committing to memory a list of facts. It includes memorization, but it also encompasses learning the underlying concepts of math that are critical to successful problem solving.

Support and Resources

Math-U-See has a number of resources to help you in the educational process.

Many of our customer service representatives have been with us for over 10 years. They are able to answer your questions, help you place your student in the appropriate level, and provide knowledgeable support throughout the school year.

Visit mathusee.com to use our many online resources, find out when we will be in your neighborhood, and connect with us on social media.

More than Memorization

Many people confuse memorization with understanding. Once while I was teaching seven junior high students, I asked how many pieces they would each receive if there were fourteen pieces. The students' response was, "What do we do: add, subtract, multiply, or divide?" Knowing how to divide is important, but understanding when to divide is equally important.

The Suggested 4-Step Math-U-See Approach

In order to train students to be confident problem solvers, here are the four steps that I suggest you use to get the most from the Math-U-See curriculum.

Step 1. Prepare for the lesson
Step 2. Present and explore the new concept together
Step 3. Practice for mastery
Step 4. Progress after mastery

Step 1. Prepare for the lesson

Watch the video lesson to learn the new concept and see how to demonstrate this concept with the manipulatives when applicable. Study the written explanations and examples in the instruction manual.

Step 2. Present and explore the new concept together

Present the new concept to your student. Have the student watch the video lesson with you, if you think it would be helpful. The following should happen interactively.

a. **Build:**   Use the manipulatives to demonstrate and model problems from the instruction manual. If you need more examples, use the appropriate lesson practice pages.

b. **Write:**   Write down the step-by-step solutions as you work through the problems together, using manipulatives.

c. **Say:**   Talk through the why of the math concept as you build and write.

Give as many opportunities for the student to "Build, Write, Say" as necessary until the student fully understands the new concept and can demonstrate it to you confidently. One of the joys of teaching is hearing a student say, *"Now I get it!"* or *"Now I see it!"*

Step 3. Practice for mastery

Using the lesson practice problems from the student workbook, have students practice the new concept until they understand it. It is one thing for students to watch someone else do a problem; it is quite another to do the same problem themselves. Together complete as many of the lesson practice pages as necessary (not all pages may be needed) until the student understands the new concept, demonstrating confident mastery of the skill. Remember, to demonstrate mastery, your student should be able to teach the concept back to you using the Build, Write, Say method. Give special attention to the word problems, which are designed to apply the concept being taught in the lesson. If your student needs more assistance, go to mathusee.com to find review tools and other resources.

Step 4. Progress after mastery

Once mastery of the new concept is demonstrated, advance to the systematic review pages for that lesson. These worksheets review the new material as well as provide practice of the math concepts previously studied. If the student struggles, reteach these concepts to maintain mastery. If students quickly demonstrate mastery, they may not need to complete all of the systematic review pages.

In the 2012 student workbook, the last systematic review page for each lesson is followed by a page called "Application and Enrichment." These pages provide a way for students to review and use their math skills in a variety of different formats. You may decide how useful these activity pages are for your particular student.

Now you are ready for the lesson tests. These were designed to be an assessment tool to help determine mastery, but they may also be used as extra worksheets. Your student will be ready for the next lesson only after demonstrating mastery of the new concept and maintaining mastery of concepts found in the systematic review worksheets.

Tell me, I forget. Show me, I understand. Let me do it, I remember.
—Ancient Proverb

To this Math-U-See adds, *"Let me teach it, and I will have achieved mastery!"*

Length of a Lesson

How long should a lesson take? This will vary from student to student and from topic to topic. You may spend a day on a new topic, or you may spend several days. There are so many factors that influence this process that it is impossible to predict the length of time from one lesson to another. I have spent three days on a lesson, and I have also invested three weeks in a lesson. This experience occurred in the same book with the same student. If you move from lesson to lesson too quickly without the student demonstrating mastery, the student will become overwhelmed and discouraged as he or she is exposed to more new material without having learned previous topics. If you move too slowly, the student may become bored and lose interest in math. I believe that as you regularly spend time working along with the student, you will sense the right time to take the lesson test and progress through the book.

By following the four steps outlined above, you will have a much greater opportunity to succeed. Math must be taught sequentially, as it builds line upon line and precept upon precept on previously-learned material. I hope you will try this methodology and move at the student's pace. As you do, I think you will be helping to create a confident problem solver who enjoys the study of math.

Rectangles, Factors, and Product
Solving for an Unknown

The word *rectangle* means right angle. It comes from the Latin word "rectus" that means right. Other languages have similar words with the same meaning. A *right angle* is a square corner. A closed shape with four square corners is a rectangle. This piece of paper is a rectangle. Have the student look for other examples.

Notice that opposite sides of a rectangle have the same length.

A *square* is a special kind of rectangle. Since a square has four right angles, it is also a rectangle. Since the four sides are the same length, it can also be classified as a square. Rectangles and squares have two dimensions because we have to measure them in two directions. These two measurements will be called the *up dimension* and the *over dimension.*

Example 1

The over dimension is three, and the up dimension is two.

A rectangle also has area, shown by the squares drawn inside it. The over and up dimensions tell how long the sides or edges are. The *area* tells how many squares are inside the rectangle. In the rectangle above, there are six squares.

Example 2

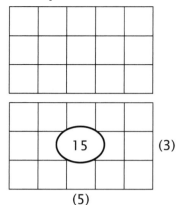

(3)

(5)

The over dimension is five, and the up dimension is three.

The area is 15 square units.

When we taught multiplication, we used rectangles to illustrate multiplication problems. We called the dimensions the *factors,* and the area was the **product.** Example 1 is a picture of 3 × 2 = 6, and Example 2 illustrates 5 × 3 = 15. When multiplying, we are given the factors, and we have to find the product. In division, we will be given the product and one factor, and we will be asked to find the missing factor. This is the same as solving for an unknown, which we did while learning our multiplication facts in the *Gamma* book.

Solving for an Unknown

The way you say the equation is key to understanding it. The problem 2X = 12 can be read as, "Two counted a certain number of times is 12" or "What number counted two times is 12?" because the **Commutative Property of Multiplication** states that you can change the order of the factors without changing the result. You might also verbalize this as "Two times what equals 12?" or "What times two equals 12?" or "Two somethings equals 12." Any of these ways is acceptable. Choose the one that makes it easiest for the student to understand.

Example 3 illustrates solving for the unknown with the manipulative blocks. Study these examples until this important concept is understood.

Example 3
Solve for the unknown, or find the value of "X" in 2X = 12.

6

2X = 12 or (2)(6) = 12

I solved it as, "How many 2s can I count out of 12?" or "What times 2 equals 12?" Using the blocks, you can see that there are six 2s in 12. Thus, 2 × 6 = 12, and the missing factor is 6. When you find the answer, simply write a 6 just above the X. The student workbook may use other letters for the unknown instead of X.

Word Problem Tips

Parents often find it challenging to teach children how to solve word problems. Here are some suggestions for helping your student learn this important skill.

The first step is to realize that word problems require both reading and math comprehension. Don't expect a child to be able to solve a word problem if he does not thoroughly understand the math concepts involved. On the other hand, a student may have a math skill level that is stronger than his or her reading comprehension skills. Below are a number of strategies to improve comprehension skills in the context of story problems. You may decide which ones work best for you and your student.

Strategies for word problems:

1. Ignore numbers at first and read the story. It may help some students to read the question aloud. Every word problem tells a story. Before deciding what math operation is required, let the student retell the story in his own words. Who is involved? Are they receiving gifts, losing something, or dividing a treat?

2. Relate the story to real life, perhaps by using names of family members or friends. For some students, this makes the problem more interesting and relevant.

3. Build, draw, or act out the story. Use the blocks or actual objects when practical. Especially in the lower levels, you may require the student to use the blocks for word problems, even when the facts have been learned. Don't be afraid to use a little drama as well. The purpose is to make it as real and meaningful as possible.

4. Look for the common language used in a particular kind of problem. Pay close attention to the word problems on the lesson practice pages, as they model different kinds of language that may be used for the new concept just studied. For example, "altogether" usually indicates addition. These "key words" can be useful clues but should not be substitutes for understanding.

5. Look for practical applications that use the concept and ask questions in that context.

6. Have the students invent word problems to illustrate their number problems from the lesson.

Cautions

1. Unneeded information may be included in the problem. For example, we may be told that Suzie is eight years old, but the eight is irrelevant when adding up the number of gifts she received.

2. Some problems may require more than one step to solve. Model these questions carefully.

3. There may be more than one way to solve some problems. Experience will help the student choose the easier or preferred method.

4. Estimation is a valuable tool for checking an answer. If an answer is unreasonable, it is possible that the wrong method was used to solve the problem.

Division by 1 and 2, Symbols for Division

Once the student grasps that division is looking for the missing factor, the division facts should be a review of his or her knowledge of multiplication. Please make sure that the multiplication facts are mastered before proceeding further.

The next things the student must become familiar with are the symbols for division. Two of the three symbols are introduced in this lesson. The first is a horizontal line as shown below, and the second is the symbol, "÷."

Figure 1

$$\frac{6}{2}$$

The first symbol is a line between the numbers to be divided.

$6 \div 2$

You can think of the division sign as related to the horizontal bar shown above. The dots above and below the line remind us where the numbers 6 and 2 used to be.

The most common way to verbalize the problem is, "Six divided by two." To enhance understanding, consider, "How many twos can I count out of six?" or "How many groups of two can I count out of six?"

Another option is to relate division to solving for the unknown, as in the problem $2X = 6$. There we read it as, "Two times what number is the same as six?" Once the student can divide by two, turn this around to $6 = 2X$ and read it as "Six is two times what number?" The answer is $6 \div 2 = 3$.

In $2X = 6$, the product and one factor are given, and we have to find the missing factor. This is the definition of division. Solving for the unknown helps us to connect division and multiplication. In multiplication, we are given the two factors and asked to find the product. In division we are given the product and one factor and asked to find the missing factor. Division is the inverse of multiplication. Notice that you cannot divide by zero because there is no number times zero that will equal any number except zero.

The division facts to be learned in this lesson are easy, so focus on the symbols and how to verbalize the problems as summarized in Example 1.

Example 1

$$\frac{6}{2}$$

$6 \div 2$

1. "What times two is equal to six?"
2. "Two times what is equal to six?"
3. "How many twos can I count out of six?"
4. "Six divided by two equals what number?"

This is a good time to show all of the division facts on a chart. After the student learns a set of facts, color or circle them. This is a good way to keep track of progress and encourage the student. There will be a small chart in each of the lessons in the instruction manual to show what has been learned. There is a chart for the student after the lesson 2 worksheets in the student workbook.

1 ÷ 1	2 ÷ 2	3 ÷ 3	4 ÷ 4	5 ÷ 5	6 ÷ 6	7 ÷ 7	8 ÷ 8	9 ÷ 9	10 ÷ 10
2 ÷ 1	4 ÷ 2	6 ÷ 3	8 ÷ 4	10 ÷ 5	12 ÷ 6	14 ÷ 7	16 ÷ 8	18 ÷ 9	20 ÷ 10
3 ÷ 1	6 ÷ 2	9 ÷ 3	12 ÷ 4	15 ÷ 5	18 ÷ 6	21 ÷ 7	24 ÷ 8	27 ÷ 9	30 ÷ 10
4 ÷ 1	8 ÷ 2	12 ÷ 3	16 ÷ 4	20 ÷ 5	24 ÷ 6	28 ÷ 7	32 ÷ 8	36 ÷ 9	40 ÷ 10
5 ÷ 1	10 ÷ 2	15 ÷ 3	20 ÷ 4	25 ÷ 5	30 ÷ 6	35 ÷ 7	40 ÷ 8	45 ÷ 9	50 ÷ 10
6 ÷ 1	12 ÷ 2	18 ÷ 3	24 ÷ 4	30 ÷ 5	36 ÷ 6	42 ÷ 7	48 ÷ 8	54 ÷ 9	60 ÷ 10
7 ÷ 1	14 ÷ 2	21 ÷ 3	28 ÷ 4	35 ÷ 5	42 ÷ 6	49 ÷ 7	56 ÷ 8	63 ÷ 9	70 ÷ 10
8 ÷ 1	16 ÷ 2	24 ÷ 3	32 ÷ 4	40 ÷ 5	48 ÷ 6	56 ÷ 7	64 ÷ 8	72 ÷ 9	80 ÷ 10
9 ÷ 1	18 ÷ 2	27 ÷ 3	36 ÷ 4	45 ÷ 5	54 ÷ 6	63 ÷ 7	72 ÷ 8	81 ÷ 9	90 ÷ 10
10 ÷ 1	20 ÷ 2	30 ÷ 3	40 ÷ 4	50 ÷ 5	60 ÷ 6	70 ÷ 7	80 ÷ 8	90 ÷ 9	100 ÷ 10

Division Facts Sheet

$1 \div 1$	$2 \div 2$	$3 \div 3$	$4 \div 4$	$5 \div 5$	$6 \div 6$	$7 \div 7$	$8 \div 8$	$9 \div 9$	$10 \div 10$
$2 \div 1$	$4 \div 2$	$6 \div 3$	$8 \div 4$	$10 \div 5$	$12 \div 6$	$14 \div 7$	$16 \div 8$	$18 \div 9$	$20 \div 10$
$3 \div 1$	$6 \div 2$	$9 \div 3$	$12 \div 4$	$15 \div 5$	$18 \div 6$	$21 \div 7$	$24 \div 8$	$27 \div 9$	$30 \div 10$
$4 \div 1$	$8 \div 2$	$12 \div 3$	$16 \div 4$	$20 \div 5$	$24 \div 6$	$28 \div 7$	$32 \div 8$	$36 \div 9$	$40 \div 10$
$5 \div 1$	$10 \div 2$	$15 \div 3$	$20 \div 4$	$25 \div 5$	$30 \div 6$	$35 \div 7$	$40 \div 8$	$45 \div 9$	$50 \div 10$
$6 \div 1$	$12 \div 2$	$18 \div 3$	$24 \div 4$	$30 \div 5$	$36 \div 6$	$42 \div 7$	$48 \div 8$	$54 \div 9$	$60 \div 10$
$7 \div 1$	$14 \div 2$	$21 \div 3$	$28 \div 4$	$35 \div 5$	$42 \div 6$	$49 \div 7$	$56 \div 8$	$63 \div 9$	$70 \div 10$
$8 \div 1$	$16 \div 2$	$24 \div 3$	$32 \div 4$	$40 \div 5$	$48 \div 6$	$56 \div 7$	$64 \div 8$	$72 \div 9$	$80 \div 10$
$9 \div 1$	$18 \div 2$	$27 \div 3$	$36 \div 4$	$45 \div 5$	$54 \div 6$	$63 \div 7$	$72 \div 8$	$81 \div 9$	$90 \div 10$
$10 \div 1$	$20 \div 2$	$30 \div 3$	$40 \div 4$	$50 \div 5$	$60 \div 6$	$70 \div 7$	$80 \div 8$	$90 \div 9$	$100 \div 10$

LESSON 3

Division by 10, Third Symbol for Division

In this lesson, we will focus on learning the ten facts. The third symbol of division looks like a half rectangle (⌐). Some call it a "house" or a "box", but it is generally known as the long division symbol. It is very effective in assisting us in relating division to multiplication.

In Figure 1 below, the product is 40, and one factor is 10. We need to find the missing factor. Remember the rectangle reviewed in lesson 1, which was used extensively when teaching multiplication. The factors are the outside dimensions, and the product is the area of the rectangle.

When we build rectangles, we often refer to the up factor and the over factor. Using this wording and drawing arrows may be helpful for the student.

Figure 1

$$10 \overline{)\,40\,}^{\,?} \qquad 10\uparrow 40 \,^{?} \qquad 40 \div 10 \qquad \frac{40}{10}$$

All of the problems in Figure 1 are the same but expressed with different symbols. They may also be verbalized several different ways:

"What times 10 is equal to 40?"
"10 times what is equal to 40?"
"How many 10s can I count out of 40?"
"40 divided by 10 equals what number?"

The rectangle in Figure 2 shows us that the answer is four. Using the blocks, you can build all the 10 facts.

Figure 2

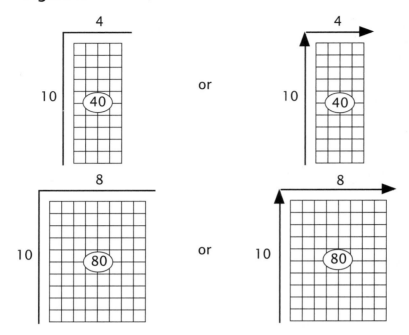

Division isn't commutative; the problem 40 ÷ 10 is very different than the problem 10 ÷ 40. However, multiplication is commutative, and every division problem may be expressed as a multiplication problem. In Figure 3, the problem is written two ways. The product is the same in each case, but the factors have changed positions. We see that $4 \times 10 = 40$ and $10 \times 4 = 40$. Both problems have the same product. When you turn the rectangle, it still has the same factors and product, or the identical dimensions and area.

Another way to express this is "40 is 10 times greater than what number?" The equation for this is $40 = 10X$. This has been solved using a student's knowledge of multiplication. Using division, we can solve it as $40 \div 10 = 4$; 40 is 10 times greater than four.

Figure 3

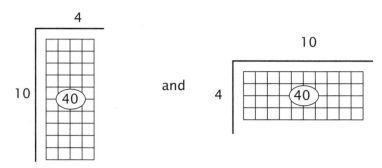

The chart shows that, when you learn the 10 facts on the far right, such as $10 \div 10 = 1$ and $20 \div 10 = 2$, you are also learning the facts on the bottom line that have 10 as a factor, such as $10 \div 1 = 10$ and $20 \div 2 = 10$.

1 ÷ 1	2 ÷ 2	3 ÷ 3	4 ÷ 4	5 ÷ 5	6 ÷ 6	7 ÷ 7	8 ÷ 8	9 ÷ 9	10 ÷ 10
2 ÷ 1	4 ÷ 2	6 ÷ 3	8 ÷ 4	10 ÷ 5	12 ÷ 6	14 ÷ 7	16 ÷ 8	18 ÷ 9	20 ÷ 10
3 ÷ 1	6 ÷ 2	9 ÷ 3	12 ÷ 4	15 ÷ 5	18 ÷ 6	21 ÷ 7	24 ÷ 8	27 ÷ 9	30 ÷ 10
4 ÷ 1	8 ÷ 2	12 ÷ 3	16 ÷ 4	20 ÷ 5	24 ÷ 6	28 ÷ 7	32 ÷ 8	36 ÷ 9	40 ÷ 10
5 ÷ 1	10 ÷ 2	15 ÷ 3	20 ÷ 4	25 ÷ 5	30 ÷ 6	35 ÷ 7	40 ÷ 8	45 ÷ 9	50 ÷ 10
6 ÷ 1	12 ÷ 2	18 ÷ 3	24 ÷ 4	30 ÷ 5	36 ÷ 6	42 ÷ 7	48 ÷ 8	54 ÷ 9	60 ÷ 10
7 ÷ 1	14 ÷ 2	21 ÷ 3	28 ÷ 4	35 ÷ 5	42 ÷ 6	49 ÷ 7	56 ÷ 8	63 ÷ 9	70 ÷ 10
8 ÷ 1	16 ÷ 2	24 ÷ 3	32 ÷ 4	40 ÷ 5	48 ÷ 6	56 ÷ 7	64 ÷ 8	72 ÷ 9	80 ÷ 10
9 ÷ 1	18 ÷ 2	27 ÷ 3	36 ÷ 4	45 ÷ 5	54 ÷ 6	63 ÷ 7	72 ÷ 8	81 ÷ 9	90 ÷ 10
10 ÷ 1	20 ÷ 2	30 ÷ 3	40 ÷ 4	50 ÷ 5	60 ÷ 6	70 ÷ 7	80 ÷ 8	90 ÷ 9	100 ÷ 10

Mastery of multiplication is essential for success in division. If you find that you need to review multiplication, go to mathusee.com for more resources that may be used to review the multiplication facts.

LESSON 4

Division by 5 and 3

We have seen from the first three lessons that division is built upon an understanding of multiplication. Make sure the student has mastered the three and five multiplication facts before working through this lesson. If you need to take time to review these facts, please do so. Math is sequential and builds upon previously learned concepts. If you need to back up and learn multiplication, consider using the Math-U-See *Gamma* level.

When I was in school, we were taught division with the expression, "Three goes into 27 how many times?" This phrase "goes into" degenerated into "gazinda." Even today, you will hear "Three gazinda 27 how many times?" Now if this reaches your ears, you will know its origin.

In a division problem, there are names for the three components. They are *divisor, dividend,* and *quotient*, as shown in Figure 1. These terms can be useful when discussing a division problem. Notice in Figure 2 how multiplication and division are related.

Figure 1

$$\text{divisor} \overline{\left| \text{dividend} \right.} \text{ quotient}$$

Figure 2

$$\text{factor} \overline{\left| \text{product} \right.} \text{ factor}$$

Learn all of the five facts and three facts. Take whatever time you need to master these before moving to the next lesson.

Example 1

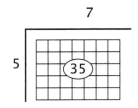

$$\frac{35}{5} = 35 \div 5 =$$

"What times five is equal to 35?"
"Five times what is equal to 35?"
"How many fives can I count out of 35?"
"35 divided by five equals what number?"

Example 2

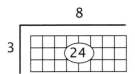

$$\frac{24}{3} = 24 \div 3 =$$

"What times three is equal to 24?"
"Three times what is equal to 24?"
"How many threes can I count out of 24?"
"24 divided by three equals what number?"

You may want to have students rewrite some of their division word problems as multiplication problems that require them to solve for the unknown. Example 2 above can be written as 3W = 24. It would be solved as 24 ÷ 3 = 8. Be sure the student knows that 3W = 24 and 24 = 3W represent the same relationship.

1 ÷ 1	2 ÷ 2	3 ÷ 3	4 ÷ 4	5 ÷ 5	6 ÷ 6	7 ÷ 7	8 ÷ 8	9 ÷ 9	10 ÷ 10
2 ÷ 1	4 ÷ 2	6 ÷ 3	8 ÷ 4	10 ÷ 5	12 ÷ 6	14 ÷ 7	16 ÷ 8	18 ÷ 9	20 ÷ 10
3 ÷ 1	6 ÷ 2	9 ÷ 3	12 ÷ 4	15 ÷ 5	18 ÷ 6	21 ÷ 7	24 ÷ 8	27 ÷ 9	30 ÷ 10
4 ÷ 1	8 ÷ 2	12 ÷ 3	16 ÷ 4	20 ÷ 5	24 ÷ 6	28 ÷ 7	32 ÷ 8	36 ÷ 9	40 ÷ 10
5 ÷ 1	10 ÷ 2	15 ÷ 3	20 ÷ 4	25 ÷ 5	30 ÷ 6	35 ÷ 7	40 ÷ 8	45 ÷ 9	50 ÷ 10
6 ÷ 1	12 ÷ 2	18 ÷ 3	24 ÷ 4	30 ÷ 5	36 ÷ 6	42 ÷ 7	48 ÷ 8	54 ÷ 9	60 ÷ 10
7 ÷ 1	14 ÷ 2	21 ÷ 3	28 ÷ 4	35 ÷ 5	42 ÷ 6	49 ÷ 7	56 ÷ 8	63 ÷ 9	70 ÷ 10
8 ÷ 1	16 ÷ 2	24 ÷ 3	32 ÷ 4	40 ÷ 5	48 ÷ 6	56 ÷ 7	64 ÷ 8	72 ÷ 9	80 ÷ 10
9 ÷ 1	18 ÷ 2	27 ÷ 3	36 ÷ 4	45 ÷ 5	54 ÷ 6	63 ÷ 7	72 ÷ 8	81 ÷ 9	90 ÷ 10
10 ÷ 1	20 ÷ 2	30 ÷ 3	40 ÷ 4	50 ÷ 5	60 ÷ 6	70 ÷ 7	80 ÷ 8	90 ÷ 9	100 ÷ 10

Parallel and Perpendicular Lines, Angles

Parallel Lines

A plane is an imaginary figure that can be represented by any flat surface, such as a piece of paper, the floor, or a desktop. It is two-dimensional because it has length and width, but it has no thickness. Straight lines that lie in the same plane and never cross or intersect are said to be *parallel.* Because they never touch, they are always the same distance apart.

It is possible to have two lines in three-dimensional space that never intersect and are not parallel. If you take two lines represented by yardsticks, it is possible to position them in such a way that they never would intersect, even if the lines represented by them were extended at both ends. These are called skew lines.

Our discussion involves lines in the same plane that never touch. A *line* is indicated by arrows at both ends to represent that it goes on indefinitely in both directions. If a line on the page doesn't have the arrows, it is only a line segment. Examples of parallel lines are the horizontal lines on a piece of lined notebook paper or the rails of perfectly straight railroad tracks. Look for other examples.

The illustrations in Figure 1 represent parallel lines. The symbol for parallel is||. Notice that the two "*ls*" in the middle of the word *parallel* look like the symbol.

Figure 1

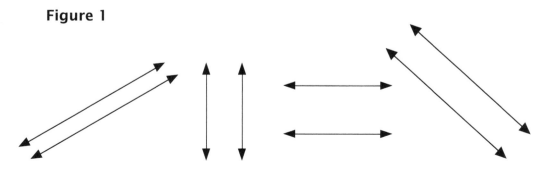

Perpendicular Lines

Two straight lines that lie in the same plane and intersect to form a *right angle* are *perpendicular lines.* At this level, you may define a right angle as a square corner. To show that the lines form a right angle, we put a little square where they meet. This means the two lines are perpendicular.

Two roads in a town can be perpendicular at an intersection if they form a right angle. An addition sign is made of two perpendicular line segments. A simple kite frame is also made up of two perpendicular line segments. The illustrations in Figure 2 represent perpendicular lines. The symbol for perpendicular is ⊥.

Figure 2

Angles

A *ray* starts at one point and goes on forever in a straight line. We use a dot to show the point where the ray starts and an arrow on the other end to show that it keeps on going. An *angle* lies between two rays that have a common starting point. Angles are measured with *degrees*. Picture two rays that have a common endpoint lying one on top of the other, like the hands on a clock. As the top ray is slowly moved in one direction, degrees measure how far it has moved. If the top ray moves 360°, it has turned a full circle and is back on top of the bottom ray.

A 90° angle is the same as the right angle we discussed above. An angle that has a measure less than 90° is called an *acute angle,* and an angle with a measure greater than 90° is an *obtuse angle.*

Using a *protractor* to draw angles is a useful activity. Follow the directions on the next page to draw a 50° angle.

Use a protractor to draw an angle with measure 50°.

1. Make a dot and draw a straight line with one endpoint (a ray). Use a ruler or anything with a straight edge.

2. Place your protractor on the ray so that the dot shows through the little hole (with the cross hairs on the endpoint). The ray should be going through 0° and 180° on one side of the protractor.

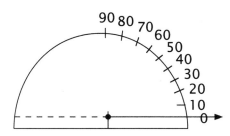

3. Beginning with 0° on the ray, count up to 50° and put a mark on your paper by 50°.

4. Remove the protractor. Then, using your ruler or straightedge, connect the vertex (common endpoint of the two rays) with the mark at 50°. Make the ray as long as you like.

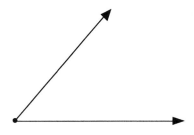

You have drawn a 50° angle. The rays that make the angle may be as long or as short as you wish. Their length will not affect the measure of the angle itself.

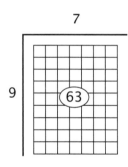
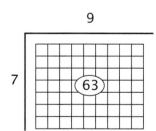

LESSON 6

Division by 9

A number is divisible by nine if the digits add up to nine or a multiple of nine. For example, $3 \times 9 = 27$, and $2 + 7 = 9$.

In this lesson, our focus is on learning all of the nine facts. Monitor the student's mastery and proceed to the next lesson only when you are satisfied with his or her progress.

Example 1

$$9 \underset{63}{\overset{?}{\longrightarrow}} \qquad \frac{63}{9} = 63 \div 9 =$$

"What times nine is equal to 63?"
"Nine times what is equal to 63?"
"How many nines can I count out of 63?"
"63 divided by nine equals what number?"

Example 2

$$7 \underset{63}{\overset{?}{\longrightarrow}} \qquad \frac{63}{7} = 63 \div 7 =$$

"What times seven is equal to 63?"
"Seven times what is equal to 63?"
"How many sevens can I count out of 63?"
"63 divided by seven equals what number?"

1 ÷ 1	2 ÷ 2	3 ÷ 3	4 ÷ 4	5 ÷ 5	6 ÷ 6	7 ÷ 7	8 ÷ 8	9 ÷ 9	10 ÷ 10
2 ÷ 1	4 ÷ 2	6 ÷ 3	8 ÷ 4	10 ÷ 5	12 ÷ 6	14 ÷ 7	16 ÷ 8	18 ÷ 9	20 ÷ 10
3 ÷ 1	6 ÷ 2	9 ÷ 3	12 ÷ 4	15 ÷ 5	18 ÷ 6	21 ÷ 7	24 ÷ 8	27 ÷ 9	30 ÷ 10
4 ÷ 1	8 ÷ 2	12 ÷ 3	16 ÷ 4	20 ÷ 5	24 ÷ 6	28 ÷ 7	32 ÷ 8	36 ÷ 9	40 ÷ 10
5 ÷ 1	10 ÷ 2	15 ÷ 3	20 ÷ 4	25 ÷ 5	30 ÷ 6	35 ÷ 7	40 ÷ 8	45 ÷ 9	50 ÷ 10
6 ÷ 1	12 ÷ 2	18 ÷ 3	24 ÷ 4	30 ÷ 5	36 ÷ 6	42 ÷ 7	48 ÷ 8	54 ÷ 9	60 ÷ 10
7 ÷ 1	14 ÷ 2	21 ÷ 3	28 ÷ 4	35 ÷ 5	42 ÷ 6	49 ÷ 7	56 ÷ 8	63 ÷ 9	70 ÷ 10
8 ÷ 1	16 ÷ 2	24 ÷ 3	32 ÷ 4	40 ÷ 5	48 ÷ 6	56 ÷ 7	64 ÷ 8	72 ÷ 9	80 ÷ 10
9 ÷ 1	18 ÷ 2	27 ÷ 3	36 ÷ 4	45 ÷ 5	54 ÷ 6	63 ÷ 7	72 ÷ 8	81 ÷ 9	90 ÷ 10
10 ÷ 1	20 ÷ 2	30 ÷ 3	40 ÷ 4	50 ÷ 5	60 ÷ 6	70 ÷ 7	80 ÷ 8	90 ÷ 9	100 ÷ 10

LESSON 7

Finding the Area of a Parallelogram

To find the area of a rectangle, multiply the base (over dimension) times the height (up dimension). We will use this simple formula or a derivative of it to find the area of other shapes. The **height** is shown by the letter *h* and is always perpendicular to the **base**. We show this with the little square in the corner where the sides of the rectangle meet or intersect.

The sides of a figure are measured in units, while the area is measured in square units. Using the blocks can help you visualize the squares inside the rectangle (Example 1). Each square represents one square foot. I combine the words "square" and "area" to form "squarea." This reminds me of the relationship between a square and area.

Example 1
Find the area of the rectangle.

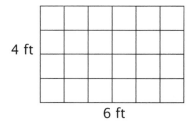

Area = bh = 6 ft × 4 ft = 24 square feet

The formula for finding the area of a *parallelogram* is the same as the formula for a rectangle. Look at Figure 1 on the next page. Notice how the triangle on the right is moved to form a rectangle. Notice also that the base and height remain the same in all the pictures.

Figure 1

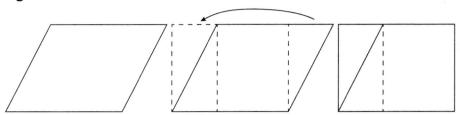

Example 2
Find the area of the parallelogram.

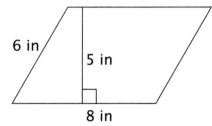

Area = bh = 8 in × 5 in = 40 square inches

Notice that the height in Example 2 is five inches and not six inches because the height is always perpendicular to the base.

Example 3
Find the area of the parallelogram.

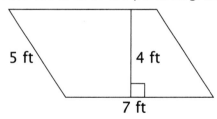

Area = bh = 7 ft × 4 ft = 28 square feet

The base can be measured either on the top or on the bottom because the two sides are the same length.

LESSON 8

Division by 6

Notice that all the multiples of six are even numbers. Notice also that the digits of the multiples also add up to three or a multiple of three. In $6 \times 7 = 42$, 42 is an even number, and $4 + 2 = 6$, which is a multiple of three. Carefully observe the student's progress and move to the next lesson only when you are satisfied with his or her mastery of the concepts.

Example 1

$$6 \overset{?}{\underset{24}{\uparrow}} \quad\quad \frac{24}{6} = 24 \div 6 =$$

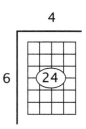

"What times six is equal to 24?"
"Six times what is equal to 24?"
"How many sixes can I count out of 24?"
"24 divided by six equals what number?"

Example 2

$$4 \overset{?}{\underset{24}{\uparrow}} \quad\quad \frac{24}{4} = 24 \div 4 =$$

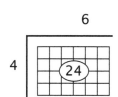

"What times four is equal to 24?"
"Four times what is equal to 24?"
"How many fours can I count out of 24?"
"24 divided by four equals what number?"

1 ÷ 1	2 ÷ 2	3 ÷ 3	4 ÷ 4	5 ÷ 5	6 ÷ 6	7 ÷ 7	8 ÷ 8	9 ÷ 9	10 ÷ 10
2 ÷ 1	4 ÷ 2	6 ÷ 3	8 ÷ 4	10 ÷ 5	12 ÷ 6	14 ÷ 7	16 ÷ 8	18 ÷ 9	20 ÷ 10
3 ÷ 1	6 ÷ 2	9 ÷ 3	12 ÷ 4	15 ÷ 5	18 ÷ 6	21 ÷ 7	24 ÷ 8	27 ÷ 9	30 ÷ 10
4 ÷ 1	8 ÷ 2	12 ÷ 3	16 ÷ 4	20 ÷ 5	24 ÷ 6	28 ÷ 7	32 ÷ 8	36 ÷ 9	40 ÷ 10
5 ÷ 1	10 ÷ 2	15 ÷ 3	20 ÷ 4	25 ÷ 5	30 ÷ 6	35 ÷ 7	40 ÷ 8	45 ÷ 9	50 ÷ 10
6 ÷ 1	12 ÷ 2	18 ÷ 3	24 ÷ 4	30 ÷ 5	36 ÷ 6	42 ÷ 7	48 ÷ 8	54 ÷ 9	60 ÷ 10
7 ÷ 1	14 ÷ 2	21 ÷ 3	28 ÷ 4	35 ÷ 5	42 ÷ 6	49 ÷ 7	56 ÷ 8	63 ÷ 9	70 ÷ 10
8 ÷ 1	16 ÷ 2	24 ÷ 3	32 ÷ 4	40 ÷ 5	48 ÷ 6	56 ÷ 7	64 ÷ 8	72 ÷ 9	80 ÷ 10
9 ÷ 1	18 ÷ 2	27 ÷ 3	36 ÷ 4	45 ÷ 5	54 ÷ 6	63 ÷ 7	72 ÷ 8	81 ÷ 9	90 ÷ 10
10 ÷ 1	20 ÷ 2	30 ÷ 3	40 ÷ 4	50 ÷ 5	60 ÷ 6	70 ÷ 7	80 ÷ 8	90 ÷ 9	100 ÷ 10

Finding the Area of a Triangle

Remember that the formula for finding the area of a rectangle or a parallelogram is base times height. The height is shown by the letter *h* and is always perpendicular to the base. Notice that the height in Example 2 is five inches and not six inches.

Example 1
Find the area of the rectangle.

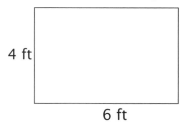

Area = bh = 6 ft × 4 ft
= 24 square feet (sq ft)

Example 2
Find the area of the parallelogram.

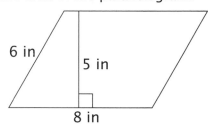

Area = bh = 8 in × 5 in
= 40 square inches (sq in)

A *triangle* is half of a rectangle or parallelogram. The formula for the area of a triangle is one half the area of a rectangle or parallelogram. In other words, it is one half the base times the height, or (½)(b)(h).

The formula can also be written as base times height divided by two: $\frac{b \times h}{2}$ or $\frac{bh}{2}$. Taking one half of a number is the same as dividing the number by two. Use the formula that works better for you.

Example 3
Find the area of the unshaded triangle.

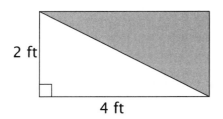

$$\text{Area} = \frac{1}{2}bh = \frac{1}{2}(4 \times 2)$$
$$= \frac{1}{2}(8) = 4 \text{ sq ft}$$
$$\text{or Area} = \frac{bh}{2} = (4 \times 2) \div 2$$
$$= (8 \div 2) = 4 \text{ sq ft}$$

Example 4
Find the area of the unshaded triangle.

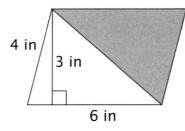

$$\text{Area} = \frac{1}{2}bh = \frac{1}{2}(6 \times 3)$$
$$= \frac{1}{2}(18) = 9 \text{ sq in}$$
$$\text{or Area} = \frac{bh}{2} = (6 \times 3) \div 2$$
$$= (18 \div 2) = 9 \text{ sq in}$$

Example 5
Find the area of the triangle.

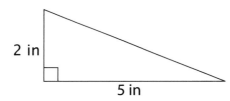

$$\text{Area} = \frac{1}{2}bh = \frac{1}{2}(5 \times 2)$$
$$= \frac{1}{2}(10) = 5 \text{ sq in}$$
$$\text{or Area} = \frac{bh}{2} = (5 \times 2) \div 2$$
$$= (10 \div 2) = 5 \text{ sq in}$$

Example 6
Find the area of the triangle.

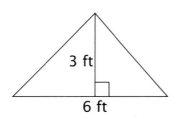

$$\text{Area} = \frac{1}{2}bh = \frac{1}{2}(6 \times 3)$$
$$= \frac{1}{2}(18) = 9 \text{ sq ft}$$
$$\text{or Area} = \frac{bh}{2} = (6 \times 3) \div 2$$
$$= (18 \div 2) = 9 \text{ sq ft}$$

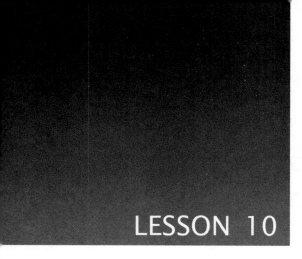

LESSON 10

Division by 4

There aren't many strategies for learning the four facts. We don't need many, since there are only three new facts that we haven't been exposed to already. The rest of the four facts are inverses of the facts we have already learned. If you learn the new ones quickly, move on to the next lesson.

In the problems with the box, have the student write the missing factor on the top. Then multiply, write the product under the number in the box, and subtract, as shown in Figure 1. We're getting ready for more difficult problems.

Figure 1

$$
4\overline{)\,24\,}^{?} \quad \rightarrow \quad 4\overline{)\,24\,}^{6} \quad \rightarrow \quad
\begin{array}{r}
6 \\
4\overline{)\,24\,} \\
\underline{-24} \\
0
\end{array}
$$

Example 1

$$
4\overset{?}{\underset{32}{\big\uparrow}}\!\!\!\longrightarrow \qquad \frac{32}{4} = 32 \div 4 = \qquad
$$

"What times four is equal to 32?"
"Four times what is equal to 32?"
"How many fours can I count out of 32?"
"32 divided by four equals what number?"

Example 2

$$4 \overset{?}{\underset{}{\uparrow}} 28 \qquad \frac{28}{4} = 28 \div 4 = \qquad 4\overline{\smash{)}\,\underset{}{(28)}} \;^{7}$$

"What times four is equal to 28?"
"Four times what is equal to 28?"
"How many fours can I count out of 28?"
"28 divided by four equals what number?"

1 ÷ 1	2 ÷ 2	3 ÷ 3	4 ÷ 4	5 ÷ 5	6 ÷ 6	7 ÷ 7	8 ÷ 8	9 ÷ 9	10 ÷ 10
2 ÷ 1	4 ÷ 2	6 ÷ 3	8 ÷ 4	10 ÷ 5	12 ÷ 6	14 ÷ 7	16 ÷ 8	18 ÷ 9	20 ÷ 10
3 ÷ 1	6 ÷ 2	9 ÷ 3	12 ÷ 4	15 ÷ 5	18 ÷ 6	21 ÷ 7	24 ÷ 8	27 ÷ 9	30 ÷ 10
4 ÷ 1	8 ÷ 2	12 ÷ 3	16 ÷ 4	20 ÷ 5	24 ÷ 6	28 ÷ 7	32 ÷ 8	36 ÷ 9	40 ÷ 10
5 ÷ 1	10 ÷ 2	15 ÷ 3	20 ÷ 4	25 ÷ 5	30 ÷ 6	35 ÷ 7	40 ÷ 8	45 ÷ 9	50 ÷ 10
6 ÷ 1	12 ÷ 2	18 ÷ 3	24 ÷ 4	30 ÷ 5	36 ÷ 6	42 ÷ 7	48 ÷ 8	54 ÷ 9	60 ÷ 10
7 ÷ 1	14 ÷ 2	21 ÷ 3	28 ÷ 4	35 ÷ 5	42 ÷ 6	49 ÷ 7	56 ÷ 8	63 ÷ 9	70 ÷ 10
8 ÷ 1	16 ÷ 2	24 ÷ 3	32 ÷ 4	40 ÷ 5	48 ÷ 6	56 ÷ 7	64 ÷ 8	72 ÷ 9	80 ÷ 10
9 ÷ 1	18 ÷ 2	27 ÷ 3	36 ÷ 4	45 ÷ 5	54 ÷ 6	63 ÷ 7	72 ÷ 8	81 ÷ 9	90 ÷ 10
10 ÷ 1	20 ÷ 2	30 ÷ 3	40 ÷ 4	50 ÷ 5	60 ÷ 6	70 ÷ 7	80 ÷ 8	90 ÷ 9	100 ÷ 10

LESSON 11

Finding the Average

The *average* is a number that represents an entire group of numbers. It is found by adding several pieces of data and then dividing by the number of pieces of data in the group. If I want to find the average age of three children who are 2, 3, and 10 years old, I start by using the blocks to represent each of the ages.

Example 1

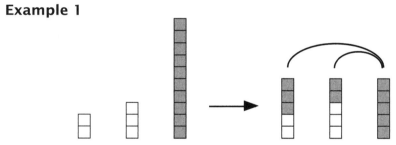

First I break up the 10 into smaller pieces using the two, three, and five unit bars. Then I move blocks from the greater age onto the lesser ages until they are all the same height. The low values are filled in, and the high value is leveled off.

The average age of the children is five. To do the same problem without blocks, add the ages of the children and divide by the total number of children, as shown in Example 2.

Example 2

2 + 3 + 10 = 15   15 ÷ 3 = 5

Example 3

During the past four months, we received the following inches of rainfall:
3 in, 7 in, 9 in, and 5 in.

What was the average amount of rain per month?

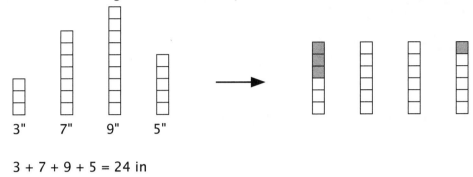

3"   7"   9"   5"

3 + 7 + 9 + 5 = 24 in

24 (total inches) ÷ 4 (number of months) = 6 inches per month

Statistics is the branch of mathematics that collects and organizes data in order to find patterns and make predictions. The skill of finding the average is an important part of statistics.

LESSON 12

Division by 7 and 8

These are the last facts to be learned. There are four multiplication facts for which you have not learned the inverses. They are $7 \times 7 = 49$, $7 \times 8 = 56$, $8 \times 7 = 56$, and $8 \times 8 = 64$. When you learn the inverses of these, you will know all of the single-digit division facts. Then you will be ready for multiple-digit division.

Continue to write out the answers to the problems in the box, as shown in Figure 1.

Figure 1

$$
\begin{array}{r} ? \\ 8\overline{\smash{)}56} \end{array}
\quad\rightarrow\quad
\begin{array}{r} 7 \\ 8\overline{\smash{)}56} \end{array}
\quad\rightarrow\quad
\begin{array}{r} 7 \\ 8\overline{\smash{)}5\ 6} \\ -5\ 6 \\ \hline 0 \end{array}
$$

Example 1

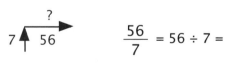

$$\frac{49}{7} = 49 \div 7 =$$

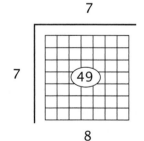

Example 2

$$\frac{56}{7} = 56 \div 7 =$$

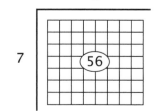

Example 3

$$8 \uparrow \overset{?}{\underset{64}{\quad}}\qquad \frac{64}{8} = 64 \div 8 = \qquad 8 \boxed{\quad 64 \quad}$$

1 ÷ 1	2 ÷ 2	3 ÷ 3	4 ÷ 4	5 ÷ 5	6 ÷ 6	7 ÷ 7	8 ÷ 8	9 ÷ 9	10 ÷ 10
2 ÷ 1	4 ÷ 2	6 ÷ 3	8 ÷ 4	10 ÷ 5	12 ÷ 6	14 ÷ 7	16 ÷ 8	18 ÷ 9	20 ÷ 10
3 ÷ 1	6 ÷ 2	9 ÷ 3	12 ÷ 4	15 ÷ 5	18 ÷ 6	21 ÷ 7	24 ÷ 8	27 ÷ 9	30 ÷ 10
4 ÷ 1	8 ÷ 2	12 ÷ 3	16 ÷ 4	20 ÷ 5	24 ÷ 6	28 ÷ 7	32 ÷ 8	36 ÷ 9	40 ÷ 10
5 ÷ 1	10 ÷ 2	15 ÷ 3	20 ÷ 4	25 ÷ 5	30 ÷ 6	35 ÷ 7	40 ÷ 8	45 ÷ 9	50 ÷ 10
6 ÷ 1	12 ÷ 2	18 ÷ 3	24 ÷ 4	30 ÷ 5	36 ÷ 6	42 ÷ 7	48 ÷ 8	54 ÷ 9	60 ÷ 10
7 ÷ 1	14 ÷ 2	21 ÷ 3	28 ÷ 4	35 ÷ 5	42 ÷ 6	49 ÷ 7	56 ÷ 8	63 ÷ 9	70 ÷ 10
8 ÷ 1	16 ÷ 2	24 ÷ 3	32 ÷ 4	40 ÷ 5	48 ÷ 6	56 ÷ 7	64 ÷ 8	72 ÷ 9	80 ÷ 10
9 ÷ 1	18 ÷ 2	27 ÷ 3	36 ÷ 4	45 ÷ 5	54 ÷ 6	63 ÷ 7	72 ÷ 8	81 ÷ 9	90 ÷ 10
10 ÷ 1	20 ÷ 2	30 ÷ 3	40 ÷ 4	50 ÷ 5	60 ÷ 6	70 ÷ 7	80 ÷ 8	90 ÷ 9	100 ÷ 10

Mental Math

These problems can be used to keep the facts fresh in the memory and to develop mental math skills. Begin slowly and walk the student through increasingly difficult exercises. The purpose is to stretch but not discourage. You decide where that line is for your student.

Example 4

"Two plus three, times four, equals what number?"

Read the problem one part at a time, waiting for the student to verbalize each step. The student thinks, "Two plus three equals five, and five times four equals twenty." As skills increase, the student should be able to do the intermediate steps mentally and give the final answer aloud.

The questions are intended to be read aloud in the order written. Do not worry about order of operations. These questions include addition, subtraction, and multiplication. Division will be included in lesson 18.

You will find more suggested mental math problems for you to read aloud to your student in lessons 18 and 24. Try a few at a time, remembering to go quite slowly at first. Be sure your student is comfortable with the shorter problems before trying the longer ones.

1. Two plus six, minus five, equals what number? (3)

2. Seven minus four, times nine, equals what number? (27)

3. Two times four, minus six, equals what number? (2)

4. Seven plus six, minus four, equals what number? (9)

5. Four plus three, times seven, equals what number? (49)

6. Two times three, minus one, plus eight, equals what number? (13)

7. Five plus seven, minus six, times seven, equals what number? (42)

8. Seventeen minus nine, plus one, times six, equals what number? (54)

9. Seven plus three, times four, plus three, equals what number? (43)

10. Three times three, minus seven, times eight, equals what number? (16)

Finding the Area of a Trapezoid

A *trapezoid* is a *quadrilateral* (four-sided figure) with two parallel sides called *bases.* The formula for the area of a trapezoid is derived from the formula for the area of a rectangle. Multiply base times height to find the area of a rectangle. Multiply the average base times the height to find the area of a trapezoid.

Figure 1

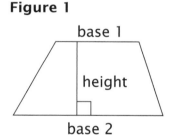

Figure 2

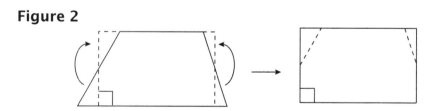

Figure 2 shows how a trapezoid is related to a rectangle. On each side of the trapezoid, find the point in the middle and make a small triangle. Pivot on the midpoints and swing the triangle up on both sides to make a rectangle out of the trapezoid. The resulting base is the average of the top and bottom bases and is found by connecting the two midpoints.

The traditional formula for finding the area of a trapezoid is $\dfrac{b_1 + b_2}{2} \times h$.

The average base is found by adding the lengths of the top and bottom bases and dividing the result by two.

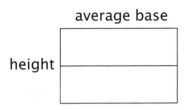

The average base length is then multiplied by the height to find the area.

Example 1
Find the area of the trapezoid.

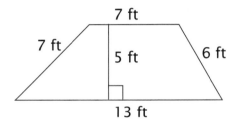

The average base length is
   $(7 + 13) \div 2 = 10$ ft.
The height is 5 ft.
The area is $10 \times 5 = 50$ sq ft.

Example 2
Find the area of the trapezoid.

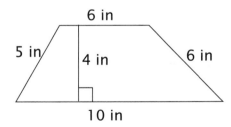

The average base length is
   $(6 + 10) \div 2 = 8$ in.
The height is 4 in.
The area is $8 \times 4 = 32$ sq in.

LESSON 14

Thousands, Millions, and Place-Value Notation

Place value is a huge component in understanding multiple-digit division. Starting on the right, the first value is units. The units place is represented by the small green cube. The next greatest place value is the tens, shown with the blue 10 bar. It is 10 times as large as a unit block. The next value to the left is the hundreds place, represented by the large red block. Notice that as you move to the left, each value is 10 times greater than the preceding value.

Figure 1

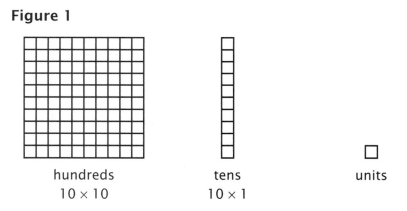

hundreds
10 × 10

tens
10 × 1

units

When you name a number such as 247, the 2 tells you how many hundreds, the 4 tells how many tens, and the 7 tells how many units. We read 247 as "Two hundred for-ty seven." (The letters "ty" mean ten, so four tens are forty.) The hundreds, tens, and units tell us what kind, or what value. The 2, 4, and 7 are digits that tell us how many. Where each digit is written, or what *place* it occupies, tells us its *value*. As the values progress from right to left, they increase by a factor of 10 because we are operating in a base 10 system, or the decimal system.

The next place value after the hundreds is the thousands place. It is 10 times 100. You could build 1,000 by stacking 10 hundred squares and making a cube. You can also show 1,000 by making a rectangle that is 10 by 100 out of the cube, as shown in figure 2. (The picture is much smaller than actual size.)

Figure 2 also shows 10,000. Can you imagine what 100,000 would look like if you use rectangles? It would be a rectangle 100 by 1,000. The factors are written inside the rectangles in the drawings.

Figure 2

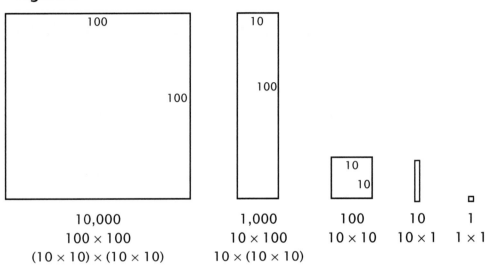

There is a lot of information in Figure 2. It shows how you multiply by 10 each time as you move from right to left. If you had room, you could draw more pictures to show that 10 times 10,000 is 100,000. Finally, 10 times 100,000 is 1,000,000 (one million).

In Figure 3, do you see that there are three places within the thousands? The same is true with the millions. There are millions, 10 millions, and 100 millions. The commas separate the number into groups of three.

Figure 3

$$\underset{\text{millions}}{1\ 2\ 3} , \underset{\text{thousands}}{4\ 5\ 6} , 7\ 8\ 9$$

When saying these greater numbers, I like to think of the commas as having names. The first comma from the right is named "thousand", and the second from the right is "million." See Example 1.

Example 1
Say 123,456,789.

"123 million 456 thousand 789" or "one hundred twenty-three million, four hundred fifty-six thousand, seven hundred eighty-nine"

Notice that you never say "and" when reading a number like this. The word "and" is reserved for the decimal point, which we will study in a later book. Practice saying and writing greater numbers.

Example 2
Say 13,762.

"13 thousand 762" or "thirteen thousand, seven hundred sixty-two"

Place-value notation is a way of writing numbers that emphasizes place value. The number given in Example 2 looks like this in place-value notation: 10,000 + 3,000 + 700 + 60 + 2. Each value is written separately.

Example 3
Write 8,543,971 with place-value notation.

8,000,000 + 500,000 + 40,000 + 3,000 + 900 + 70 + 1

LESSON 15

Billions, Trillions, and Expanded Notation
Multi-Step Word Problems

The next two place values beyond the millions are the billions and the trillions.

Figure 1

trillions	billions	millions	thousands

1 3 5 , 2 4 6 , 1 2 3 , 5 5 6 , 7 8 9

Example 1
Say: 135,246,123,456,789

"135 trillion, 246 billion, 123 million, 456 thousand 789" or "one hundred thirty-five trillion, two hundred forty-six billion, one hundred twenty-three million, four hundred fifty-six thousand, seven hundred eighty-nine"

We have already learned that, as the values progress from right to left, they increase by a factor of 10. There is another pattern that occurs as you move from right to left among the place values. It is illustrated in Figure 2 on the next page.

Look at the first number to the left of each comma. Begin with one unit and then look at one thousand, one million, one billion, and one trillion. Because there are three spaces between each arrow, each one is $10 \times 10 \times 10$, or 1,000 times as great as the preceding value. The first arrow points at 1, the next arrow to the left points at 1,000, the next arrow points at 1,000,000, and so on.

Figure 2

$$\frac{1}{\uparrow}, \quad \underline{}\ \underline{}\ \frac{1}{\uparrow}, \quad \underline{}\ \underline{}\ \frac{1}{\uparrow}, \quad \underline{}\ \underline{}\ \frac{1}{\uparrow}, \quad \underline{}\ \underline{}\ \frac{1}{\uparrow}$$

1,000,000,000,000   1,000,000,000   1,000,000   1,000   1

A thousand is a thousand units.
A million is a thousand thousands.
A billion is a thousand millions.
A trillion is a thousand billions.

Expanded Notation

Another notation to learn is *expanded notation.* It separates the digit and the place value and goes one step further than place-value notation.

Example 2
Write 43,971 in expanded notation.

$$4 \times 10,000 + 3 \times 1,000 + 9 \times 100 + 7 \times 10 + 1$$

Multi-Step Word Problems

The student workbook includes some fairly simple two-step word problems. Some students may be ready for more challenging problems. Here are a few to try, along with some tips for solving this kind of problem. You may want to read and discuss these with your student as you work out the solutions together. Again, the purpose is to stretch, not to frustrate. If you do not think the student is ready, you may want to come back to these later.

There are more multi-step word problems in lessons 21 and 27 of the instruction manual. The solutions for the multistep problems follow the test solutions at the back of this book.

1. David has a rectangular garden that measures 11 feet by 13 feet. He wants to plant peas in his garden. Dad said that one seed packet will be enough to fill a space 10 feet on each side. Will David's garden have enough space to plant two seed packets?

Although the problem asks only one question, there are other questions that must be answered first. The key to solving this is determining what the unstated questions are. Since the final question is really asking for a comparison of the available area to the needed area, the two unstated questions are: "What is the area of David's garden?" and "What is the area needed for two seed packets?"

You might make a list of steps something like this:

1. Find the area of the garden in square feet.

2. Find the area needed for one seed packet.

3. Find the area needed for two seed packets.

4. Compare the answers to parts 1 and 4.

2. Rachel and Sarah started out to visit Grandma. They drove for 50 miles and stopped to rest before driving for 30 more miles. Then they decided to go back 10 miles to a restaurant they had seen. After leaving the restaurant, they drove 80 more miles to Grandma's house. How many miles did the girls drive on the way to Grandma's house?

Make a drawing, and this will be easier to solve!

3. Rachel and Sarah spent $8 for gasoline, $15.65 for their lunch, and $5 apiece for gifts for Grandma. Grandma gave each of them $10. If the girls left home with a total of $50, how much money do they have for the return trip?

This is similar to the previous one in that you must answer other questions before you can answer the question in the problem.

.

LESSON 16

Division by a Single Digit with Remainder

So far in this course, all the answers to division problems have come out evenly. This does not happen in real life. For example, if I have 13 quarters, how many dollars do I have? There are four quarters in a dollar, but 13 ÷ 4 is not one of the division facts we have learned.

The problem can be stated as, "How many groups of four quarters are in 13 quarters?" or "13 ÷ 4 is equal to what number?" Example 1 uses numbers, symbols, and blocks to show how to work out a problem with a *remainder.*

Example 1

How many groups of 4 are in 13? The answer is 3.

Multiply: 3 x 4 =12. Place the 12 under the 13 and subtract.

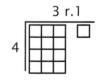

13 minus 12 leaves 1, which is the remainder. We write it as "r.1."

Therefore, if I have 13 quarters, I have three groups of 4 quarters with one quarter left over. In Example 1, the remainder cannot be greater than 3. If the remainder had been 4, you would have another group of 4 and would be able to divide by 4 another time. The blocks show that, in this case, the remainder must be a 0, 1, 2, or 3.

Problems that have a whole number answer, which you can divide evenly, have a remainder of zero. Consider the next few examples.

Example 2

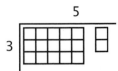
How many groups of 3 are in 17? The answer is 5.

Multiply: 5 × 3 = 15. Place the 15 under the 17 and subtract.

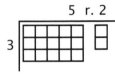
17 minus 15 leaves 2, which is the remainder. We write it as "r.2."

Example 3

How many groups of 9 are in 87? The answer is 9.

Multiply: 9 × 9 = 81. Place the 8 under the 87 and subtract.

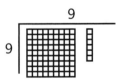
87 minus 81 leaves 6, which is the remainder. We write it as "r.6."

"Upside Down" Multiplication

The student should already know how to multiply multiple-digit numbers. This lesson teaches "upside down" multiplication. The purpose is to review the role of place value in multiplication and to prepare us for division. This method makes it easier to keep track of place value.

We'll do each of the problems with regular notation and place-value notation. When possible, we'll use the blocks as well. Example 1 is the traditional way of multiplying multiple-digit numbers, and Example 2 is the upside down way.

Example 1 (traditional)

```
   21        20 + 1
 ×  3       ×     3
 ───        ───────
   63        60 + 3
```

20   + 1

Example 1 (upside down)

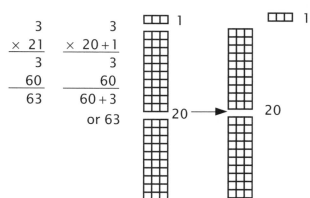

```
    3          3
 × 21       × 20 + 1
 ───        ───────
    3          3
   60         60
 ───        ───────
   63         60 + 3

            or 63
```

20

3   3

This is the same problem as above done vertically. Move the top three blocks into the units place.

Example 2

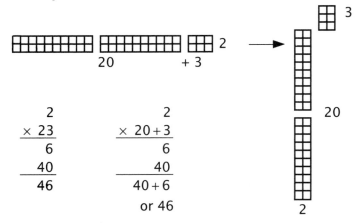

```
    2            2
  × 23        × 20+3
    6            6
   40           40
   46          40+6

            or 46
```

Example 3 ("upside down" and traditional)

```
    3            3          132    100 + 30 + 2
  × 132       × 100+30+2    ×  3   ×          3
    6            6          396    300 + 90 + 6
   90           90
  300          300
  396          396

          or 300 + 90 + 6
```

Example 4 ("upside down" and traditional)

```
    6            6          218          200+10+8
  × 218       × 200+10+8    ×  6         ×        6
   48           48          ¹4            ¹⁰⁰  40
   60           60         1268        1000+200+60+8
 1200         1200         1308        1000+300+00+8
 1308         1308
```

As you look at these division problems, watch for the pattern in the place value. In Example 5, there are 3 groups of 3 in 9, and there are 30 groups of 3 in 90. Because the 9 is in the tens place and because 9 divided by 3 is 3, you can cover up the zero, and the answer will have a 3 in the tens place.

Example 5

$$\begin{array}{r} 30 \\ 3\overline{\smash{\big)}\,90} \\ -90 \\ \hline 0 \end{array}$$

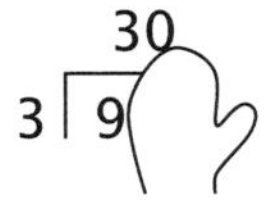

Likewise, you can cover up both of the zeros in Example 6. Looking only at the hundreds place and covering up the rest, we see $7 \div 7 = 1$. When we consider place value, we see that this means that $700 \div 7 = 100$.

Example 6

$$\begin{array}{r} 100 \\ 7\overline{\smash{\big)}\,700} \\ -700 \\ \hline 0 \end{array}$$

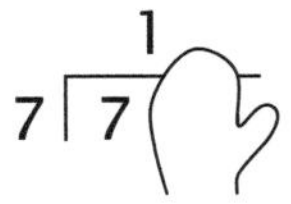

In Examples 7 and 8, we look at the first *two* digits (the ones that are not zero). In Example 7, there are 4 groups of 6 in 24, so there will be 40 groups of 6 in 240. In Example 8, we have $63 \div 9 = 7$ and $630 \div 9 = 70$.

Example 7

$$\begin{array}{r} 40 \\ 6\overline{\smash{\big)}\,240} \\ -240 \\ \hline 0 \end{array}$$

Example 8

$$\begin{array}{r} 70 \\ 9\overline{\smash{\big)}\,630} \\ -630 \\ \hline 0 \end{array}$$

We will use these patterns to break bigger division problems down into smaller parts, looking at only one or two digits at a time.

LESSON 18

Division with Double-Digit Factors

When multiplying, we multiply the units first, then the tens, then the hundreds, and so forth. When dividing, we begin by looking at the greatest place value first. This should not be surprising because division is the inverse of multiplication.

We look for the greatest number that can be multiplied to get a value close to the number under the division sign. Then we subtract and ask ourselves what can be multiplied by the divisor to get close to the new number.

In the examples, we will walk the student through each step. Notice how we work each problem with the manipulatives and then with place-value notation and regular notation.

Example 1

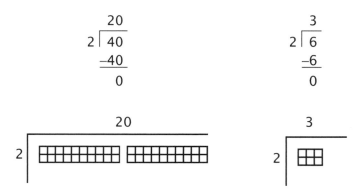

```
      20                3
   2 | 40            2 | 6
     -40              -6
       0                0
```

Example 1 shows two small problems: 40 ÷ 2 and 6 ÷ 2. In Example 2, the smaller problems are combined. First we divide the tens place and then the units place. The two-step process is new.

Example 2

How many groups of 2 are in 46?

```
        3
       20
    2 | 46
      -40
        6
       -6
        0
```

Start by looking at the digit in the tens place. We know that there are 2 groups of 2 in 4, so we also know that there will be at least 20 groups of 2 in 40.

Put a 20 on top as part of that factor. Multiply $2 \times 20 = 40$ and write 40 under the 46. Then subtract: $46 - 40$ leaves a remainder of 6.

```
       20 + 3
    2 | 40 + 6
      -40
        6
       -6
        0
```

How many groups of 2 are in 6?

There are 3 groups of 2 in 6, so we add a 3 to the top factor. Multiply $2 \times 3 = 6$ and subtract 6 from the previous remainder of 6. Because $6 - 6 = 0$, there is no further remainder.

Add the 20 and the 3 to find the missing factor. The answer to the division problem is 23.

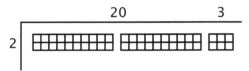

Check your answer by multiplying.

```
        2                2
   ×  20 + 3         ×  23
        6                6
       40               40
     40 + 6  or 46      46
```

Example 3

$$
\begin{array}{r}
2 \\
10 \\
4\,\overline{)\,48} \\
-40 \\
\hline
8 \\
-8 \\
\hline
0
\end{array}
$$

How many groups of 4 are in 48?

Look at the digit in the tens place. There is 1 group of 4 in 4, so there will be at least 10 groups of 4 in 40.

Put a 10 on top. Multiply $4 \times 10 = 40$ and write 40 under the 48. Subtract: 48 – 40 leaves a remainder of 8.

How many groups of 4 are in 8?

$$
\begin{array}{r}
10+2 \\
4\,\overline{)\,40+8} \\
-40 \\
\hline
8 \\
-8 \\
\hline
0
\end{array}
$$

There are 2 groups of 4 in 8, so we add a 2 to the top factor. Multiply $4 \times 2 = 8$ and subtract 8 from the previous remainder of 8.

Because 8 – 8 = 0, there is no further remainder. Add the 10 and 2 together to find the missing factor: $48 \div 4 = 12$

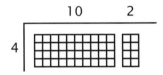

Always check your answer by multiplying.

$$
\begin{array}{rr}
4 & 4 \\
\times\ 10+2 & \times\ 12 \\
\hline
8 & 8 \\
40 & 40 \\
\hline
40+8 \text{ or } 48 & 48
\end{array}
$$

Example 4

$$\begin{array}{r} 2 \\ 30 \\ 3\overline{)98} \\ -90 \\ \hline 8 \\ -6 \\ \hline 2 \end{array}$$

How many groups of 3 are in 98?

Look at the digit in the tens place. There are 3 groups of 3 in 9, so there will be at least 30 groups of 3 in 90.

Put a 30 on top. Multiply $3 \times 30 = 90$ and write 90 under the 98. Subtract: 98 – 90 leaves a remainder of 8.

How many groups of 3 are in 8?

$$\begin{array}{r} 30+2 \\ 3\overline{)90+8} \\ -90 \\ \hline 8 \\ -6 \\ \hline 2 \end{array}$$

There are 2 groups of 3 in 8, so we add a 2 to the top factor. Multiply $3 \times 2 = 6$ and subtract 6 from the previous remainder of 8.

Because 8 – 6 = 2, we have a remainder of 2. Since we cannot evenly divide 2 into 3 parts, we leave this remainder alone for now.

$98 \div 3 = 32$ r.2

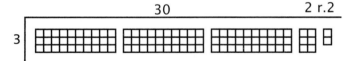

Because we weren't able to divide evenly, multiplication won't give us the original number. Multiply $3 \times 32 = 96$ and then add the remainder to get $96 + 2 = 98$.

$$\begin{array}{r} 3 \\ \times \quad 30+2 \\ \hline 6 \\ 90 \\ \hline 90+6+2=98 \end{array} \qquad \begin{array}{r} 3 \\ \times 32 \\ \hline 6 \\ 90 \\ \hline 96 \\ 2 \\ \hline 98 \end{array}$$

Example 5

How many groups of 2 are in 56?

```
        8
       20
   2 ⟌ 56
      −40
       16
      −16
        0
```

Look at the digit in the tens place. There are 2 groups of 2 in 5, so there will be at least 20 groups of 2 in 50. (There are actually more than 20 groups of 2 in 50, but we are only asking how many groups of 2 are in 5 and taking 10 times that many.)

Put a 20 on top. Multiply $2 \times 20 = 40$ and write 40 under the 56. Subtract: 56 − 40 leaves a remainder of 16.

```
      20 + 8
   2 ⟌ 50 + 6
      −40
      10 + 6
     −10 + 6
        0
```

How many groups of 2 are in 16?

There are 8 groups of 2 in 16, so we add an 8 to the top. Multiply $8 \times 2 = 16$ and subtract 16 from the previous remainder.

Because 16 − 16 = 0, there is no further remainder. Add the 20 and the 8 to get the final answer: $56 \div 2 = 28$

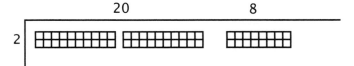

Always check your answer by multiplying.

```
          2              2
    ×   20 + 8      ×   28
          16            16
          40            40
    40 + 16 = 56        56
```

Example 6

```
      9
     10
  5 | 95
    −50
     45
    −45
      0
```

How many groups of 5 are in 95?

Look at the digit in the tens place. There is 1 group of 5 in 9, so there will be at least 10 groups of 5 in 90.

Put a 10 on top. Multiply $5 \times 10 = 50$ and write 50 under the 95. Subtract: 95 − 50 leaves a remainder of 45.

How many groups of 5 are in 45?

```
      10 + 9
  5 | 90 + 5
    −50
     40 + 5
    −40 + 5
        0
```

There are 9 groups of 5 in 45, so we add a 9 to the top factor. Multiply $5 \times 9 = 45$ and subtract 45 from the previous remainder.

Because 45 − 45 = 0, there is no further remainder. Add the 10 and the 9 to get the final answer.

$$95 \div 5 = 19$$

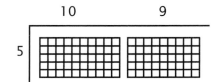

Always check your answer by multiplying.

```
       5           5
  × 10 + 9      × 19
      45          45
      50          50
      95          95
```

Mental Math

Here are some more mental math problems for you to read aloud to your student. Try a few at a time, going slowly at first. Be sure your student is comfortable with the shorter problems before trying the longer ones.

1. Forty-two divided by six, plus five, equals? (12)

2. Three times four, divided by six, equals? (2)

3. Thirteen minus five, divided by two, equals? (4)

4. Seven plus eight, divided by three, equals? (5)

5. Twenty minus ten, divided by five, equals? (2)

6. Six plus six, minus three, divided by three, equals? (3)

7. Forty-five divided by five, plus one, plus ten, equals? (20)

8. Eighteen minus nine, plus seven, divided by two, equals? (8)

9. Two times three, times four, divided by eight, equals? (3)

10. Seventy-two divided by eight, divided by three, times seven, equals? (21)

Division, Three Digit by One Digit

The question we have been asking for division is, "How many groups?" Study the examples carefully to see how to think this through step by step.

Example 1

```
       2
      10
     300
  3 | 936
    -900
      36
     -30
       6
      -6
```

How many groups of 3 are in 936?

Start in the hundreds place by looking at the 9 in 936. There are 3 groups of 3 in 9, so we can start by taking 300 groups of 3.

Put 300 on top. Multiply $3 \times 300 = 900$ and subtract 900 from 936, leaving a remainder of 36.

Look at the 3 in 36 (tens place). There is 1 group of 3 in 3, so we can take 10 groups of 3 out of 36.

Put a 10 on top. Multiply $3 \times 10 = 30$ and subtract 30 from 36, leaving a remainder of 6.

Look at the remainder of 6. There are 2 groups of 3 in 6, so we add 2 to the top, multiply $3 \times 2 = 6$, and subtract.

$6 - 6 = 0$, so there is no further remainder.

Add $300 + 10 + 2$ to find the answer.

$936 \div 3 = 312$

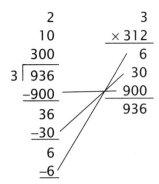

$$
\begin{array}{r}
2 \\
10 \\
300 \\
3\,\overline{)936} \\
-900 \\
\hline
36 \\
-30 \\
\hline
6 \\
-6 \\
\end{array}
\qquad
\begin{array}{r}
3 \\
\times\,312 \\
\hline
6 \\
30 \\
900 \\
\hline
936 \\
\end{array}
$$

When you check your answer with multiplication, notice how the partial products of the multiplication problem correspond to the division process.

Example 2

$$
\begin{array}{r}
8 \\
20 \\
200\ \text{r.1} \\
2\,\overline{)457} \\
-400 \\
\hline
57 \\
-40 \\
\hline
17 \\
-16 \\
\hline
1 \\
\end{array}
$$

How many groups of 2 are in 457?

Start in the hundreds place by looking at the 4 in 457. There are 2 groups of 2 in 4, so we can start by taking 200 groups of 2.

Put 200 on top. Multiply $2 \times 200 = 400$ and subtract 457 – 400 for a remainder of 57.

Next, look at the 5 in 57 (tens place). There are 2 groups of 2 in 5, so we can take 20 groups of 2 out of 57.

Add a 20 on top. Multiply $2 \times 20 = 40$ and subtract 40 from 57, leaving a remainder of 17.

Look at the remainder of 17. There are 8 groups of 2 in 17, so we add 8 to the top, multiply $2 \times 8 = 16$, and subtract.

17 – 16 = 1, so we are left with a remainder of 1.

The number 1 cannot be evenly divided by 2 (there are no groups of 2 in 1), so we leave the final remainder alone and write it with our answer.

$457 \div 2 = 228$ r.1

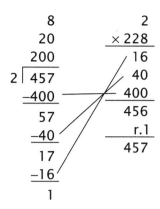

```
    8
   20
  200
2 | 457
 −400
   57
  −40
   17
  −16
    1
```

```
    2
 × 228
   16
   40
  400
  456
  r.1
  457
```

Because this number didn't divide evenly, multiplication won't give us the original number. Multiply $2 \times 228 = 456$ and then add the remainder to get $456 + 1 = 457$.

In Examples 3 and 4, we encounter two types of problems that can be difficult if you are just memorizing a formula. However, if we understand what we are doing, the procedure should be clear. Read them through carefully.

Example 3

```
    6
   10
9 | 144
 −90
  54
 −54
   0
```

How many groups of 9 are in 144?

Start in the hundreds place by looking at the 1 in 144. There are no groups of 9 in 1, so we move on to the tens place.

Look at the first two digits, which are 1 and 4. (They actually represent 140, or 14 tens.) There is 1 group of 9 in 14, so we can take 10 groups of 9 out of 144.

Add a 10 on top. Multiply $9 \times 10 = 90$ and subtract 90 from 144, leaving a remainder of 54.

Look at the remainder of 54. There are 6 groups of 9 in 54. Add 6 to the top, multiply $9 \times 6 = 54$, and subtract.

$54 - 54 = 0$. Add $10 + 6$ to find the answer: $144 \div 9 = 16$

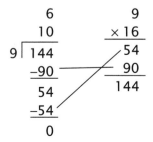

Example 4

```
        7
       00
      100 r.2
   3 ⟌ 323
     -300
       23
       -0
       23
      -21
        2
```

How many groups of 23 are in 323?

Start in the hundreds place by looking at the 3 in 323. There is 1 group of 3 in 3, so we can start by taking 100 groups of 3.

Put 100 on top. Multiply $3 \times 100 = 300$ and subtract $323 - 300$ for a remainder of 23.

Look at the 2 in the tens place. There are no groups of 3 in 2, so we move on to the units place.

Look at both digits in the remainder of 23. There are 7 groups of 3 in 23.

Add 7 to the top, multiply $3 \times 7 = 21$, and subtract. $23 - 21 = 2$, so we are left with a remainder of 2.

2 cannot be evenly divided by 3 (there are no groups of 3 in 2), so we leave the final remainder alone and write it with our answer.

$323 \div 3 = 107$ r. 2

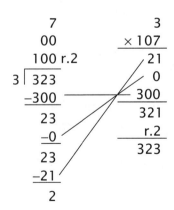

```
        3
     × 107
       21
        0
      300
      321
      r.2
      323
```

Because we weren't able to divide evenly, multiplication won't give us the original number. Multiply $3 \times 107 = 321$. Then add the remainder to get $321 + 2 = 323$.

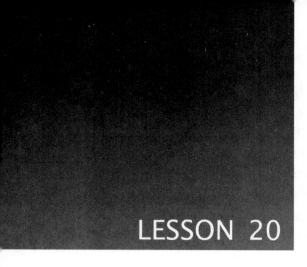

Division, Three Digit by One Digit (cont.)
Using Fractions to Divide Remainders

So far in our study of division, we have written remainders when we can't divide evenly, using the letter "r" beside the quotient. Sometimes it makes sense to write the remainder as the numerator of a fraction with the divisor of the division problem as the denominator of the fraction. In Example 1, a sample division problem is written with the "house" and as a fraction with a line between the two numbers to be divided.

Example 1

$$\frac{9}{4} = \frac{8}{4} + \frac{1}{4} = 2 + \frac{1}{4} = 2\frac{1}{4}$$

Verbalizing this problem helps us to understand the new symbolism. Nine divided by four is two, with one left over. If we divide the remainder by 4, that problem may be written as $1 \div 4$, or as $1/4$ (one fourth). We can write 2 r.1 as 2¼.

Example 2

$$\frac{33}{7} = \frac{28}{7} + \frac{5}{7} = 4 + \frac{5}{7} = 4\frac{5}{7}$$

Although we haven't formally studied fractions, we know that a line between two numbers indicates division. Think of dividing nine pies evenly among four people. Each person will get two pies, but there will be one pie left over. That pie can be divided into four pieces so that each person gets two whole pies and one fourth of a pie. The fourth of a pie is 1 pie divided by 4. We can write it as 1/4.

Study Example 3. Although we are really dividing 2 into 300 in the first step, we usually think, "How many groups of 2 can we count out of 3?" or "How many times does 2 go into 3?" For now, we will continue to write the answer above the line as a multiple of 100.

Example 3

$$
\begin{array}{r}
6 \\
70 \\
100 \\
2\,\overline{)\,353} \\
-200 \\
\hline
153 \\
-140 \\
\hline
13 \\
-12 \\
\hline
1
\end{array}
\quad \frac{1}{2}
$$

Think, "How many groups of 2 are in 3?" (See above.)

The answer is 1. Put the 1 in the hundreds place. Multiply: $2 \times 100 = 200$. Write 200 under the 353 and subtract: $353 - 200 = 153$

How many groups of 2 are in 1? We can't count any groups of 2 out of 1.

How many groups of 2 are in 15? The answer is 7.

Put 7 in the tens place. Multiply $2 \times 70 = 140$. Write 140 under the 153 and subtract: $153 - 140 = 13$.

How many groups of 2 are in 13? The answer is 6. It goes in the units place. Multiply: $2 \times 6 = 12$

Put 12 under the 13 and subtract: $13 - 12 = 1$

The remainder is 1. Divide the remainder by writing the fraction 1/2 and adding it to the quotient.

$353 \div 2 = 176\frac{1}{2}$

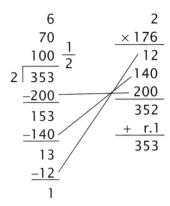

```
         6              2
        70            × 176
       100  1/2        / 12
     2 | 353        / 140
        −200 ——    − 200
         153          352
        −140        +  r.1
          13          353
         −12
           1
```

By this point, students should understand the role of place value in division. If they wish, they can begin writing answers on one line. They should continue to think about place value as they solve for the missing factor.

If a student is more comfortable with writing it all out as before, you may allow this. I still like to check by multiplying "upside down," but either way will work, as shown in Example 4.

Example 4

```
        132  3/6
     6 | 795
       −600
        195
       −180
         15
        −12
          3
```

"How many groups of 6 are in 7?" The answer is 1. Put 1 in the hundreds place.

Multiply: 6 × 100 = 600. Write 600 under the 795 and subtract: 795 − 600 = 195

How many groups of 6 are in 1? We can't count any groups of 6 out of 1.

How many groups of 6 are in 19? The answer is 3. Write 3 in the tens place. Multiply: 6 × 30 = 180. Put 180 under the 195 and subtract: 195 − 180 = 15.

How many groups of 6 are in 15? The answer is 2. It goes in the units place. Multiply: 6 × 2 = 12. Write 12 under the 15 and subtract: 15 − 12 = 3

Divide the remainder by writing the fraction 3/6 and adding it to the quotient. (Students have not yet learned to simplify fractions, so leave it as 3/6.)

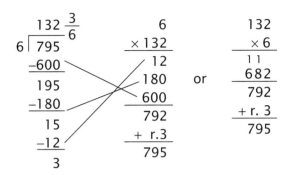

```
      3
  132 —
      6
6 | 795        6         132
  −600       × 132       × 6
  ─────      ─────       ────
   195         12        1 1
  −180        180        682
  ─────       600        ───
    15    or  ───        792
   −12        792       + r. 3
  ─────      + r.3       ───
     3        ───        795
              795
```

Example 5

```
      4
  167 —
      5
5 | 839
  −500
  ────
   339
  −300
  ────
    39
   −35
  ────
     4
```

How many groups of 5 are in 8? The answer is 1.
Write the 1 in the hundreds place.

Multiply: 5 × 100 = 500. Write 500 under the 839 and
subtract: 839 − 500 = 339

How many groups of 5 are in 3?
We can't count any groups of 5 out of 3.

How many groups of 5 are in 33? The answer is 6.

Write 6 in the tens place. Multiply: 5 × 60 = 300
Write 300 under the 339 and subtract: 339 − 300 = 39

How many groups of 5 are in 39? The answer is 7.

It goes in the units place. Multiply: 5 × 7 = 35. Write 35
under the 39 and subtract: 39 − 35 = 4

The remainder is 4. Divide the remainder by writing
the fraction 4/5 and adding it to the quotient.

```
      4
  167 —
      5
5 | 839        5         167
  −500       × 167       × 5
  ────       ─────       ────
   339         35        3 3
  −300        300        505
  ────        500        ───
    39    or  ───        835
   −35        835       + r.4
  ────       + r.4       ───
     4        ───        839
              839
```

Rounding to 10, 100, 1,000, and Estimation

Most of this lesson should be review, as we have covered this material in the earlier levels. If rounding is a new concept to the student, take time to be sure it is thoroughly understood. Rounding to 10 is used in estimating when we multiply or divide. When you round a number to the nearest multiple of 10, there will be a digit in the tens place but only a zero in the units place. We call it rounding because the units are going to be a "round zero."

Let's round 38 as an example. The first skill is to find the two multiples of 10 that are nearest to 38. The lesser multiple is 30, and the greater multiple is 40. We see that 38 is between 30 and 40. If the student has trouble finding these numbers, begin by placing your finger over the eight in the units place so that all you have is a three in the tens place. A three in the tens place is 30. Then add one more ten to find the 40. I often write the numbers 30 and 40 above the number 38 on both sides, as shown in Figure 1.

Figure 1

30   40
  38

The next skill is find out whether 38 is closer to 30 or 40. Let's go through all the numbers as shown in Figure 2. It is obvious that 31, 32, 33, and 34 are closer to 30 and 36, 37, 38, and 39 are closer to 40, but 35 is a special case. It is just as close to 30 as it is to 40. There are different rules that can be used with numbers ending in 5, but the rule we will use here is to round numbers with a 5 in the units place up to the nearest ten. When rounding to tens, look at the units place. If the units are 0, 1, 2, 3, or 4, the digit in the tens place remains unchanged. If the units are 5, 6, 7, 8, or 9, the digit in the tens place increases by one. See Figure 2.

Figure 2

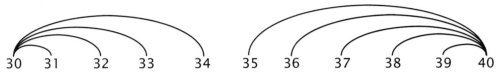

Another strategy I use is to put 0, 1, 2, 3, and 4 inside a circle to represent "0" because, if these numbers are in the units place, they add nothing to the tens place, and they are rounded to the lesser multiple of ten (30 in the example). Then I write 5, 6, 7, 8, and 9 inside a thin rectangle to represent "1" because, if these numbers are in the units place, they add one to the tens place and are rounded to the greater multiple of ten (40 in the example).

Figure 3

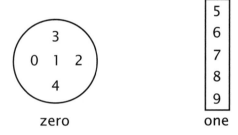

Example 1
Round 43 to the nearest tens place.

40 50 43	1. Find the multiples of 10 nearest to 43.
40 ⌐ 50 43	2. We know that three goes to the lesser multiple of ten, which is 40.
(40) ⌐ 50 43	3. Recall that three is in the circle (zero) so nothing is added to the lesser multiple of ten, which is 40.

When rounding to hundreds, look only at the digit in the tens place to determine whether to stay the same or increase by one. The same rules apply to hundreds as to tens; if the digit is a 0, 1, 2, 3, or 4, the digit in the hundreds place remains unchanged. If the digit in the tens place is a 5, 6, 7, 8, or 9, then the digit in the hundreds place increases by one. See Example 2.

Example 2
Round 547 to the nearest hundreds place.

500   600   1. Find the multiples of 100 nearest to 547.
  547

500 ⌐ 600   2. We know that four goes to the lesser
  547   multiple of 100, which is 500.

(500) ⌐ 600   3. Recall that four is in the circle (zero), so
  547   nothing is added to the lesser multiple of
                       100, which is 500.

When rounding to thousands, consider only the digit immediately to the right of the thousands place, which is the hundreds, to determine whether to stay the same or increase by one.

Example 3
Round 8,719 to the nearest thousands place.

8,000   9,000   1   Find the nearest multiples of 1,000.
  8,719

8,000 ⌐ 9,000   2. We know that seven goes to the greater
  8,719   multiple, which is 9,000.

8,000 ⌐(9,000)   3. Recall that seven is in the rectangle
  8,719   (one), so one is added to the eight.
                       The answer is 9,000.

Estimation

Now that we know how to round numbers, we can apply this skill to find the approximate answer for a division problem. In Example 4, we are only going to round the dividend. Later, when the divisor has multiple digits, we will round both of them. Round 627 to 600 first, use your hand to cover the zeros, and quickly divide three into six to get two. Then add the zeros again to get 200. Finally, complete the division and compare your approximation to the exact answer. (The symbol ≈ means "approximately equal to.")

Note: Finding the correct first number is not nearly as important as getting the correct place in which to put the first number. In other words, finding the estimated place value is the most important part of this exercise. In Example 5, the first number will not turn out to be a two in the final answer, but as long as you know it will be in the hundreds place, that is what is most critical.

Example 4
Estimate the answer.

$$3\overline{)627} \rightarrow 3\overline{)(600)} \rightarrow 3\overline{)(6)}^{2} \rightarrow 3\overline{)(600)}^{200}$$

Example 5
Estimate the answer.

$$4\overline{)781} \rightarrow 4\overline{)(800)} \rightarrow 4\overline{)(8)}^{2} \rightarrow 4\overline{)(800)}^{200}$$

Example 6
Estimate the answer.

$$2\overline{)895} \rightarrow 2\overline{)(900)} \rightarrow 2\overline{)(9)}^{4} \rightarrow 2\overline{)(900)}^{400}$$

Multi-Step Word Problems

Here are a few more multi-step word problems to try. You may want to read and discuss these with your student as you work out the solutions together. Again, the purpose is to stretch, not to frustrate. If you do not think the student is ready, you may want to come back to these later.

The solutions for the multistep problems follow the test solutions at the back of this book.

1. Scott bought 3 bags of candy with 75 pieces in each one. He plans to divide all the candy evenly among seven friends. How many pieces of candy will Scott have left for himself?

2. Anne earned $3 an hour babysitting and $4 an hour working in the garden. Last week she watched children for five hours and worked in the garden for three hours. How much more money does she need to buy a game that costs $35?

3. Riley had a nature collection. She had 25 acorns, 16 dried seed pods, and 8 feathers. She divided the acorns into five equal groups, the seed pods into four equal groups, and the feathers into two equal groups. She gave her mother one group of each kind. How many separate items did her mother get?

Using Estimation and Common Sense to Check Word Problems

Encourage students to use estimation to check the answers to word problems for reasonableness. Remind them that final answers may be somewhat different than the estimated answers. Even so, an estimate of 30 should make the student re-examine a final answer of 300.

Common sense is also a helpful tool for evaluating answers. Modeling different ways to look at a problem can be very useful to a student. In number 1 above, an answer of seven or more does not make sense because, if Scott wants to divide all his candy, he could give each friend at least one more piece. In number 2, Anne worked for a total of eight hours. At $3.00 an hour, she would earn $24.00 and need only $11.00 more to buy her game. Since she earned more than $3.00 an hour for some of her work, the actual amount she needs is less than $11.00. For number three, a student may be tempted just to add up the various items in the nature collection. Careful reading shows that Mom did not get all of the collection.

In fact, the numbers of items that Riley gave to her mother from each part of her collection is quite a bit less than the total number of items in each part.

If a student uses the incorrect operation (or operations) for a word problem, take some time to discuss why the answer may not be a sensible result. Encourage the student to think of the problem as something happening in real life, not just words on a page. You may need to model this way of thinking frequently for some students.

Division, Three Digit by Two Digit

The opposite of multiplying is dividing. The inverse of the multiplication problem $12 \times 13 = 156$ is $156 \div 13 = 12$. Here are a few examples that use the blocks and written notation to review this relationship.

Example 1
Solve: $156 \div 13$

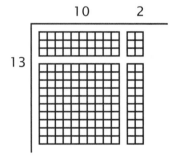

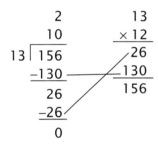

$$156 \div 13 = 12$$

Example 2
Solve: $276 \div 12$

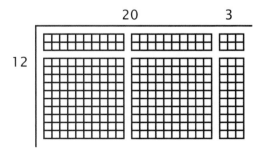

$$276 \div 12 = 23$$

Example 3
Solve: 387 ÷ 11

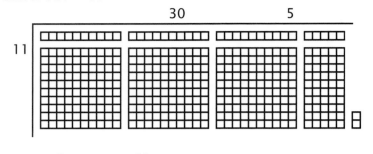

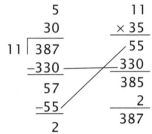

$$
\begin{array}{r}
5 \\
30 \\
11\overline{)387} \\
-330 \\
\hline
57 \\
-55 \\
\hline
2
\end{array}
\qquad
\begin{array}{r}
11 \\
\times\,35 \\
\hline
55 \\
330 \\
\hline
385 \\
2 \\
\hline
387
\end{array}
$$

387 ÷ 11 = 35 r.2

Example 4
Solve: 303 ÷ 13

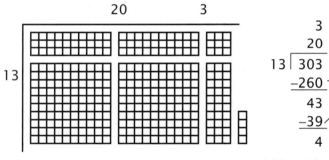

$$
\begin{array}{r}
3 \\
20 \\
13\overline{)303} \\
-260 \\
\hline
43 \\
-39 \\
\hline
4
\end{array}
\qquad
\begin{array}{r}
13 \\
\times\,23 \\
\hline
39 \\
260 \\
\hline
299 \\
4 \\
\hline
303
\end{array}
$$

303 ÷ 13 = 23 r.4

Hint: Start with three hundred blocks and three unit blocks. Break up one of the hundreds into 10 ten blocks.

Now we'll try some without the blocks and work them through carefully. Begin by estimating the answer.

Example 5

How many groups of 12 are in 7?
We can't count any 12s out of 7.

$$
\begin{array}{r}
12 \\
\times 60 \\
\hline
720
\end{array}
\qquad
\begin{array}{r}
61 \frac{11}{12} \\
12\overline{)743} \\
-720 \\
\hline
23 \\
-12 \\
\hline
11
\end{array}
\qquad
\begin{array}{r}
(70) \\
(10)\overline{)(700)}
\end{array}
$$

How many groups of 12 are in 70?
The answer is 6. It goes in the tens place.

$$
\begin{array}{r}
12 \\
\times 1 \\
\hline
12
\end{array}
$$

Multiply: $12 \times 60 = 720$.
Subtract: $743 - 720 = 23$.

How many groups of 12 are in 23?
The answer is 1. It goes in the units place.

$$
\begin{array}{r}
61 \frac{11}{12} \\
12\overline{)743} \\
-720 \\
\hline
23 \\
-12 \\
\hline
11
\end{array}
\qquad
\begin{array}{r}
12 \\
\times 61 \\
\hline
12 \\
720 \\
\hline
732 \\
+ \text{ r.}11 \\
\hline
743
\end{array}
$$

Multiply: $12 \times 1 = 12$. Subtract: $23 - 12 = 11$.
The remainder is 11.

Divide the remainder by writing the fraction 11/12 and adding it to the quotient.

$$743 \div 12 = 61^{11}/_{12}$$

When you were finding the first factor, you may have run into some difficulty. Here are some tips.

1. Use estimation. Cover the 3 with your mitten so that you have "What times 12 is 74?" You can even move the mitten over to see "What times 1 is 7?" (See Figure 1.) The answer is 7.

Figure 1

2. Experiment. If you multiply 7 times 12, you will get 84, which is too much. Try 6 and find that 6×12 is 72, and 60×12 is 720, which will work. (See Figure 2 on the next page.) It is perfectly fine to experiment until you get a useful result.

Figure 2

```
        7 0              6 0
  1 2 | 7 4 3      1 2 | 7 4 3
        8 4 0              7 2 0
  too much          just right
```

Example 6

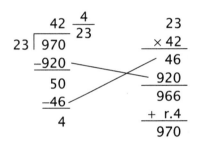

```
   23         42  4/23        (40)
  × 40    23 | 970       (20)|(900)
  920        −920
                50
   23         −46
  × 2          4
   46
```

```
          42  4/23           23
     23 | 970              × 42
       −920                  46
          50                920
       −46                  966
           4              + r.4
                            970
```

How many groups of 23 are in 97?

The answer is 4. Write it in the tens place.

Multiply: 23 × 40 = 920
Subtract: 970 − 920 = 50

How many groups of 23 are in 50?

The answer is 2. Write it in the units place.

Multiply: 23 × 2 = 46
Subtract: 50 − 46 = 4

The remainder is 4.

Divide the remainder by writing the fraction 4/23 and adding it to the quotient.

970 ÷ 23 = 42 4/23

Division, Four Digit by One Digit

In this lesson, we'll learn how to divide with greater numbers. Take your time and study the examples. When estimating, focus on the place value of the first digit in the quotient. This is more important than the actual number itself.

Example 1

$$\begin{array}{r} 9 \\ \times\ 600 \\ \hline 5,400 \end{array}$$

$$\begin{array}{r} 9 \\ \times\ 70 \\ \hline 630 \end{array}$$

$$\begin{array}{r} 9 \\ \times\ 3 \\ \hline 27 \end{array}$$

$$\begin{array}{r} 673\ \tfrac{3}{9} \\ 9\,\overline{)\,6,060} \\ -5,400 \\ \hline 660 \\ -630 \\ \hline 30 \\ -27 \\ \hline 3 \end{array}$$

$$\begin{array}{r} (600) \\ (9)\,\overline{)\,(6,000)} \end{array}$$

How many groups of 9 are in 6? We can't count any groups of 9 out of 6.

How many groups of 9 are in 60? The answer is 6. Write it in the hundreds place. Multiply: $9 \times 600 = 5,400$
Subtract: $6,060 - 5,400 = 660$.

How many groups of 9 are in 66? The answer is 7. It goes in the tens place. Multiply: $9 \times 70 = 630$
Subtract $660 - 630 = 30$

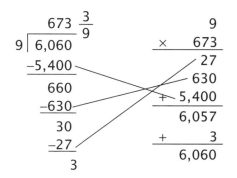

$$\begin{array}{r} 9 \\ \times\ 673 \\ \hline 27 \\ 630 \\ +\ 5,400 \\ \hline 6,057 \\ +\quad 3 \\ \hline 6,060 \end{array}$$

How many groups of 9 are in 30? The answer is 3. Multiply: $9 \times 3 = 27$
Subtract: $30 - 27 = 3$

The remainder is 3. Divide the remainder by writing the fraction 3/9 and adding it to the quotient.

Example 2

$$
\begin{array}{r}
4 \\
\times\,2{,}000 \\
\hline
8{,}000
\end{array}
\qquad
\begin{array}{r}
(2{,}000) \\
\hline
(4)\,\big|\,(9{,}000)
\end{array}
$$

$$
\begin{array}{r}
4 \\
\times\,100 \\
\hline
400
\end{array}
\qquad
\begin{array}{r}
2{,}127\,\tfrac{1}{4} \\
4\,\big|\,8{,}509 \\
-8{,}000 \\
\hline
509 \\
-400 \\
\hline
109 \\
-80 \\
\hline
29 \\
-28 \\
\hline
1
\end{array}
$$

$$
\begin{array}{r}
4 \\
\times\,20 \\
\hline
80
\end{array}
$$

How many groups of 4 are in 8?

The answer is 2. It goes in the thousands place. Multiply: $4 \times 2{,}000 = 8{,}000$
Subtract: $8{,}509 - 8{,}000 = 509$

How many groups of 4 are in 5?

The answer is 1. It goes in the hundreds place. Multiply: $4 \times 100 = 400$
Subtract: $509 - 400 = 109$

How many groups of 4 are in 10?

The answer is 2. It goes in the tens place. Multiply: $4 \times 20 = 80$
Subtract: $109 - 80 = 29$

How many groups of 4 are in 29?

The answer is 7. Multiply: $4 \times 7 = 28$
Subtract: $29 - 28 = 1$

The remainder is 1. Divide the remainder by writing the fraction 1/4 and adding it to the quotient.

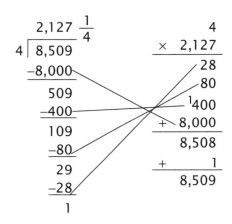

$$
\begin{array}{r}
4 \\
\times\,2{,}127 \\
\hline
28 \\
80 \\
{}^{1}400 \\
+\;8{,}000 \\
\hline
8{,}508 \\
+\quad\;\;1 \\
\hline
8{,}509
\end{array}
$$

LESSON 24

Division, Four Digit by Two Digit

In this lesson, we'll learn how to divide greater numbers by double-digit factors. This is just one more step in enabling the student to solve any division problem.

Example 1

$$
\begin{array}{r}
12 \\
\times\,300 \\
\hline
3{,}600
\end{array}
\qquad
\begin{array}{r}
(500) \\
(10)\,\overline{)(5{,}000)}
\end{array}
$$

$$
\begin{array}{r}
12 \\
\times\,90 \\
\hline
1 \\
980 \\
\hline
1{,}080
\end{array}
\qquad
\begin{array}{r}
395\,\tfrac{11}{12} \\
12\,\overline{)4{,}751} \\
-3{,}600 \\
\hline
1{,}151 \\
-1{,}080 \\
\hline
71 \\
-60 \\
\hline
11
\end{array}
$$

$$
\begin{array}{r}
12 \\
\times\,5 \\
\hline
1 \\
50 \\
\hline
60
\end{array}
$$

How many groups of 12 are in 40?
The answer is 3. Put it in the hundreds place.
Multiply: $12 \times 300 = 3{,}600$.
Subtract: $4{,}751 - 3{,}600 = 1{,}151$.

How many groups of 12 are in 11?
We can't count any groups of 12 out of 11.

How many groups of 12 are in 115?
The answer is 9. Put it in the tens place.
Multiply: $12 \times 90 = 1{,}080$.
Subtract: $1{,}151 - 1{,}080 = 71$.

How many groups of 12 are in 71?
The answer is 5. Multiply: $12 \times 5 = 60$.
Subtract: $71 - 60 = 11$.

The remainder is 11.

Divide the remainder by writing the fraction 11/12 and adding it to the quotient.

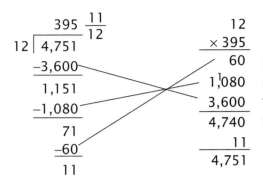

$$395\frac{11}{12}$$
$$12\overline{)4{,}751}$$
$$-3{,}600$$
$$\overline{1{,}151}$$
$$-1{,}080$$
$$\overline{71}$$
$$-60$$
$$\overline{11}$$

```
         12
      × 395
      ─────
         60
      1,080
      3,600
      ─────
      4,740
         11
      ─────
      4,751
```

Because we already worked the smaller multiplication problems out to the left of the problem, we took some shortcuts here.

Example 2

```
     38
  × 200
  ─────
  8,000
```

$$(200)$$
$$(400)\overline{)(9{,}000)}$$

```
              238 19/38
         38 ) 9,063
             −7,600
             ──────
              1,463
             −1,140
             ──────
                323
               −304
             ──────
                 19
```

Estimating is a good place to begin to find the first factor.

How many groups of 38 are in 9,000? Try the estimated 200.
Multiply: 38 × 200 = 7,600.
Subtract: 9,063 − 7,600 = 1,463.

```
     38
  × 30
  ─────
  1,140
```

How many groups of 38 are in 14? We can't count any groups of 38 out of 14. How many groups of 38 are in 140? The answer is 3. Put 3 in the tens place. Multiply: 38 × 30 = 1,140. Subtract: 1,463 − 1,140 = 323.

```
     38
  × 8
  ─────
   304
```

How many groups of 38 are in 323? The answer is 8.
Multiply: 38 × 8 = 304.
Subtract: 323 − 304 = 19.
Divide the remainder by writing the fraction 19/38 and adding it to the quotient.

$$238\frac{19}{38}$$
$$38\overline{)9{,}063}$$
$$-7{,}600$$
$$\overline{1{,}463}$$
$$-1{,}140$$
$$\overline{323}$$
$$-304$$
$$\overline{19}$$

```
        38
     × 238
     ─────
       304
     1,140
     7,600
     ─────
     9,044
        19
     ─────
     9,063
```

Symmetry

Throughout this book, students have recognized shapes by angles and the presence of parallel or perpendicular lines. Another way of classifying shapes is to determine whether or not they have some kind of symmetry.

In this lesson, we introduce symmetry around a line. Figures that are symmetrical around a line form mirror images on both sides of a line drawn through the middle of the figure. Below are two examples of figures that are symmetrical around a line, as well as one that is not symmetrical around a line.

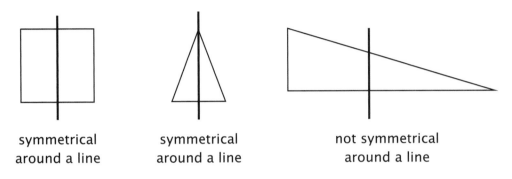

| symmetrical around a line | symmetrical around a line | not symmetrical around a line |

Mental Math

Here are some more mental math problems for you to read aloud to your student. You may shorten these if the student is not quite ready for them.

1. Twenty-five divided by five, plus three, times seven, equals? (56)

2. Sixteen minus nine, times two, plus one, equals? (15)

3. Two plus three, times four, divided by two, equals? (10)

4. Five times six, divided by ten, times six? (18)

5. Twenty-five minus one, divided by eight, plus four, equals? (7)

6. Seventy-two divided by nine, minus one, times six, equals? (42)

7. Forty-four plus one, divided by five, divided by three, equals? (3)

8. Six minus five, plus eleven, divided by six, equals? (2)

9. Two times seven, plus two, divided by two, equals? (8)

10. Seven times seven, minus one, divided by eight, times five, equals? (30)

Division, Multiple Digit by Multiple Digit

When working out a division problem, many people write the products from each step without the final zeros. Students may use this method if they understand that the position of the numbers makes the place value clear without the additional zeros. In Examples 1 and 2, I show both ways of writing the problems. Notice that, after you subtract, you bring down only one additional number (or one place value) before dividing again.

Example 1

$$
\begin{array}{r}
53 \\
\times\,500 \\
\hline
1 \\
25,500 \\
\hline
26,500
\end{array}
$$

$$
\begin{array}{r}
53 \\
\times\,60 \\
\hline
1 \\
3,080 \\
\hline
3,180
\end{array}
$$

$$
\begin{array}{r}
53 \\
\times\,8 \\
\hline
2 \\
404 \\
\hline
424
\end{array}
$$

$$
(50)\overline{)(30,000)}\,(600)
$$

$$
\begin{array}{r}
568\ \tfrac{5}{53} \\
53\,\overline{)\,30,109} \\
-26,500 \\
\hline
3,609 \\
-3,180 \\
\hline
429 \\
-424 \\
\hline
5
\end{array}
$$

How many groups of 53 are in 30? We can't count any groups of 53 out of 30. How many groups of 53 are in 300? Our estimate shows 6, but when we multiply, we find 6 is too large. The answer is 5. It goes in the hundreds place.
Multiply: 53 × 500 = 26,500.
Subtract: 30,109 – 26,500 = 3,609.
How many groups of 53 are in 360? The answer is 6. It goes in the tens place. Multiply: 53 × 60 = 3,180.
Subtract: 3,609 – 3,180 = 429.
How many groups of 53 are in 429? The answer is 8.
Multiply: 53 × 8 = 424.
Subtract: 429 – 424 = 5
The remainder may be written as 5/53.

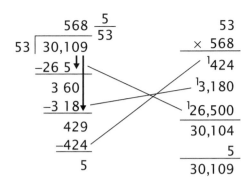

$$
\begin{array}{r}
568 \ \frac{5}{53} \\
53\overline{)30{,}109} \\
-26\ 5\downarrow \\
3\ 60 \\
-3\ 18\downarrow \\
429 \\
-424 \\
5
\end{array}
\qquad
\begin{array}{r}
53 \\
\times\ 568 \\
\hline
{}^{1}424 \\
{}^{1}3{,}180 \\
{}^{1}26{,}500 \\
\hline
30{,}104 \\
5 \\
\hline
30{,}109
\end{array}
$$

Because we already worked the smaller multiplication problems out to the left of the problem, we took some shortcuts here.

Example 2

$$
\begin{array}{r}
136 \\
\times\ 4{,}000 \\
\hline
1\ 2 \\
424{,}000 \\
\hline
544{,}000
\end{array}
$$

$$
\begin{array}{r}
136 \\
\times\ 200 \\
\hline
1 \\
26{,}200 \\
\hline
27{,}200
\end{array}
$$

$$
\begin{array}{rr}
136 & 136 \\
\times\ 50 & \times\ 7 \\
\hline
1\ 3 & 2\ 4 \\
5{,}500 & 712 \\
\hline
6{,}800 & 952
\end{array}
$$

$$
\begin{array}{r}
(6{,}000) \\
(100)\overline{)(600{,}000)}
\end{array}
$$

$$
\begin{array}{r}
4{,}257\ \frac{10}{136} \\
136\overline{)578{,}962} \\
-544{,}000 \\
\hline
34{,}962 \\
-27{,}200 \\
\hline
7{,}762 \\
-6{,}800 \\
\hline
962 \\
-952 \\
\hline
10
\end{array}
$$

How many groups of 136 are in 578? Try 6. No, too big. Try 5. No, still too large. The answer is 4. It goes in the thousands place. Multiply: $136 \times 4{,}000 = 544{,}000$. Subtract: $578{,}962 - 544{,}000 = 34{,}962$.

How many groups of 136 are in 349? The answer is 2. It goes in the hundreds place. Multiply: $136 \times 200 = 27{,}200$. Subtract $34{,}962 - 27{,}200 = 7{,}762$.

How many groups of 136 are in 776? The answer is 5. It goes in the tens place. Multiply: $136 \times 50 = 6{,}800$. Subtract: $7{,}762 - 6{,}800 = 962$.

How many groups of 136 are in 962? The answer is 7. Multiply: $136 \times 7 = 952$. Subtract: $962 - 952 = 10$. The remainder may be written as 10/136.

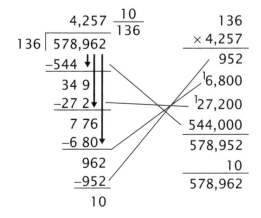

$$
\begin{array}{r}
4{,}257\ \frac{10}{136} \\
136\overline{)578{,}962} \\
-544\downarrow \\
34\ 9 \\
-27\ 2\downarrow \\
7\ 76 \\
-6\ 80\downarrow \\
962 \\
-952 \\
\hline
10
\end{array}
\qquad
\begin{array}{r}
136 \\
\times\ 4{,}257 \\
\hline
952 \\
{}^{1}6{,}800 \\
{}^{1}27{,}200 \\
544{,}000 \\
\hline
578{,}952 \\
10 \\
\hline
578{,}962
\end{array}
$$

LESSON 26

Volume

Area tells us how many square units can fit inside a two-dimensional figure. *Volume* tells us how many cubic units can fit inside a three-dimensional shape. The shapes in this lesson all have sides that are squares or rectangles. The sides are perpendicular to each other. These shapes may be called *rectangular solids* or *rectangular prisms*.

When doing these problems, I relate the shape to a hotel. First, we find the number of rooms on the ground floor (area of the base) and then multiply that by the number of floors in the hotel (the height). The formula for volume is the area of the base times the height, or V = Bh, where the capital B represents the area of the base and h represents the height. As always, the height is perpendicular to the base. You will also see the formula given as V = lwh, or length times width times height. Either way of writing the problem gives the same result.

In the first few problems *unit cubes* (cubes with a side length of one unit) are clearly shown, so have the students build the problem with a stack of blocks. After this is done, write the dimensions of the solid. In subsequent problems, the dimensions will be given, as in Example 3, but the unit cubes will not be shown. Of course, the student may draw them in if that would be helpful. Remember, the units of measure in a volume problem are cubic because we multiplied three units.

Example 1
Find the volume.

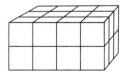

V = Bh
V = (4 units × 3 units) × 2 units
V = 24 cubic units

Example 2
Find the volume.

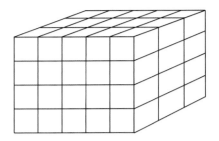

V = Bh
V = (5 units × 3 units) × 4 units
V = 60 cubic units

Example 3
Find the volume. Be sure to label your answer with the given units.

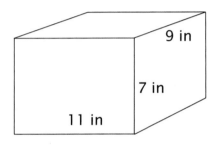

V = Bh
V = (11 in × 9 in) × 7 in
V = 693 cubic inches (cu in)

Volume and Weight of Water

In the student workbook, there are problems that ask the amount of water in a cubic foot. There are about 7.48 gallons of water in a cubic foot, and the weight of a gallon is about 8.345 pounds. Since decimals have not been introduced yet, we will round the gallons to seven gallons per cubic foot and the weight to eight pounds per gallon. These rounded values are useful when all you need is an estimated or rough answer.

Fraction of a Number

This lesson introduces three steps to understanding fractions. They will be the basis for much of our future study of fractions.

The first step is very important because it determines our starting point. A *fraction* is a "fraction of" something. This lesson stresses this fact. I often ask students, "Which is larger, one half or one fourth?" They usually reply, "One half," when what they should say is "One half of what?" and "One fourth of what?" If I then say, "One half of this room or one fourth of the state?" the answer will be different. Therefore, the first step is determing the beginning number.

The second step is determining the *denominator*. This is the number of equal parts into which we divide the starting number. Notice that both "divide" and "denominator" begin with a *d*. The symbolism is also an indicator of this step, since the line separating the numerator and denominator means "divided by." We are dividing the amount from step one by the bottom number of the fraction (the denominator). One equal part is indicated by the numeral 1 written over the denominator. You may hear this referred to as a unit fracton.

In the third step, we count how many of the equal parts are indicated by the numerator. I call this the "numBerator" at first to make the connection with counting. After a while, we take out the *B* and have *numerator*. Here is a summary of the steps.

1. The starting number; in Example 1, this is six.
2. The denominator indicates the **value** (as in place value) and tells into how many equal parts we divide the starting number.
3. The "numBerator" or numerator tells **how many** of these equal parts we count.

Example 1

Find 2/3 of 6.

$$6 \div 3 = 2$$
$$2 \times 2 = 4$$

$$\frac{2}{3} \text{ of } 6 = 4$$

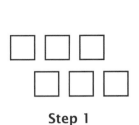

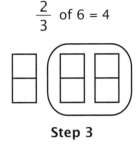

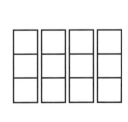

Step 1

Select 6 green unit blocks.

Step 2

Each part is 1/3 (one third) of the original number.

Step 3

Together, the parts are 2/3 (two thirds) of the original number.

A fraction is a combination of a division problem and a multiplication problem. First, we divide by the denominator, and then we multiply by the numerator. In Example 1, we divided 6 by 3 to find the equal parts. ($6 \div 3 = 2$) Then we multiplied the result by 2 to find the amount that is 2/3 of 6. ($2 \times 2 = 4$)

Example 2

Find 3/4 of 12.

$$\frac{3}{4} \text{ of } 12 = 9$$

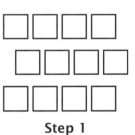

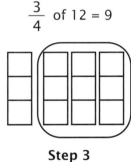

Step 1

Select 12 green unit blocks.

Step 2

Each part is 1/4 (one fourth) of the original number.

Step 3

Together, the parts are 3/4 (three fourths) of the original number.

Multi-Step Word Problems

Here are a few more multi-step word problems to try. You may want to read and discuss these with your student as you work out the solutions together. Remember that the purpose is to stretch, not to frustrate.

The solutions for the multi-step problems follow the test solutions at the back of this book.

1. Sixty-five bags of nuts are to be divided equally among 13 students. Each bag contains 15 nuts. How many nuts will each student receive?

2. Shane is playing a board game. For his first turn, he moved ahead three spaces; for the second, five spaces; and for the third, one space. For his next turn, he had to go back six spaces. After that he got a card that said he could move two times the biggest forward move he had made so far. Now how many spaces from the beginning is Shane's game piece?

3. The volume of a rectangular box is 330 cubic inches. The length on one side of the top is 11 inches, and the height of the box is 3 inches. What is the area of the top of the box?

   (A drawing may help you with this one.)

4. Tom divided $360 equally among his six children. His daughter Kate added $20 to her portion and then used the money to buy 16 gifts that each cost the same amount. What was the cost of each of Kate's gifts?

Roman Numerals: I, V, X, L, and C

Many years ago, the Romans used number symbols that resembled letters of the alphabet. This system, commonly referred to as Roman numerals, spread throughout the lands of the old Roman Empire and was used throughout Europe during the Middle Ages. That is why we still see it used today in Western countries. You may see Roman numerals used for dates, clock faces, points in an outline, and the names of kings.

Roman numerals use a base-ten system, since they describe numbers using multiples of ten and one hundred (as well as half tens and half hundreds). However, the value of a number is not based on its position in the Roman system, and there is no symbol for zero. This makes computation with Roman numerals more difficult.

After the Middle Ages, Roman numerals were largely replaced by the Hindu-Arabic number system that we use today. The symbols used in the Hindu-Arabic system are sometimes referred to simply as "Arabic numerals." As you know, this system does include zero, and the position of the symbols indicates place value.

In this lesson we will introduce five of the symbols used to write Roman numerals. Notice that they are all capital (uppercase) letters.

Three of the symbols are powers of ten. This means that we can multiply the value of each symbol by ten to get the value of the next symbol in the list. We will call symbols that are powers of ten the "primary symbols."

$$I = 1 \qquad X = 10 \qquad C = 100$$

The other two symbols we are introducing in this lesson will be referred to as "secondary symbols." They represent halves of the powers of ten.

V = 5   L = 50

Here are three basic rules for reading Roman numerals.

Rule 1   Symbols which are repeated are added together.

III = 3   XX = 20

Rule 2   Symbols with greater values to the left of symbols with lesser values are also added.

XV = 15   VII = 7

Rule 3   A single symbol with a lesser value to the left of a symbol with a greater value is subtracted from the total instead of being added.

IV = 4   IX = 9

Here are the Roman numerals for numbers 1 through 40.

1	I	11	XI	21	XXI	31	XXXI
2	II	12	XII	22	XXII	32	XXXII
3	III	13	XIII	23	XXIII	33	XXXIII
4	IV	14	XIV	24	XXIV	34	XXXIV
5	V	15	XV	25	XXV	35	XXXV
6	VI	16	XVI	26	XXVI	36	XXXVI
7	VII	17	XVII	27	XXVII	37	XXXVII
8	VIII	18	XVIII	28	XXVIII	38	XXXVIII
9	IX	19	XIX	29	XXIX	39	XXXIX
10	X	20	XX	30	XXX	40	XL

Here are some more examples to study.

50	L	80	LXXX	150	CL	300	CCC
60	LX	90	XC	200	CC	350	CCCL
70	LXX	100	C	250	CCL		

Example 1
Write CXCIII using Arabic numerals.

C is 100, XC is 100 – 10, or 90 (rule 3), and III is 3, so the answer is 100 + 90 + 3 = 193.

Example 2
Write CCCLXXIX using Arabic numerals.

CCC is 300, L is 50, XX is 20, and IX is 9 (rule 3), so the answer is 300 + 50 + 20 + 9 = 379.

There are also some commonly followed conventions or principles for writing Roman numerals. These are of more modern origin, and you may see exceptions. In real life, students are more likely to have to read than to write Roman numerals. Although these conventions are not as important to master as the rules above, they are still interesting to know. Some of the exercises in the student workbook do require students to write numbers with Roman numerals.

Convention 1   The primary symbols (I, X, and C) may be repeated. The secondary symbols (V and L) are not repeated.

XXIII for 23 is correct, but VVV for 15 is unlikely.

Convention 2   The primary symbols are commonly subtracted, but the secondary symbols are generally not subtracted.

XC for 90 is correct, but VC for 95 is unlikely.

Convention 3   The same symbol is seldom repeated more than three times.

IV is commonly used instead of IIII for 4, and XC is used rather than LXXXX for 90. An exception is a clock face, where you will often see IIII used for 4.

Convention 4   Primary symbols that have the same place value or one place value below symbols of greater value are usually the only ones to be subtracted.

IX, IV, XC, and XL are common. IC and IL are uncommon.

When writing a Roman numeral, it is helpful to write the Arabic numeral with place-value notation and then follow the rules and conventions to write each part of the number. Once you have determined the Roman numeral for each part, string them together to write the complete number.

Example 3
Show 168 with Roman numerals.

168 is 100 + 60 + 8, which is C + LX + VIII or CLXVIII.

Example 4
Show 249 with Roman numerals.

249 is 200 + 40 + 9, which is CC + XL + IX or CCXLIX.

LESSON 29

Fraction of One

In lesson 27, we learned three steps for finding the fraction of a number. Now we will apply the same steps to finding a fraction of one.

Step 1 is to note that the square in Example 1 represents one whole. Step 2 counts how many equal pieces the square is divided into. The square is divided into five equal parts. (This relates to the written problem where the number below the line is 5, and the line means "divided by.") Step 3 counts the number of shaded parts . The number on top of the line tells us two of the five equal parts are shaded.

Example 1
Find 2/5 of 1.

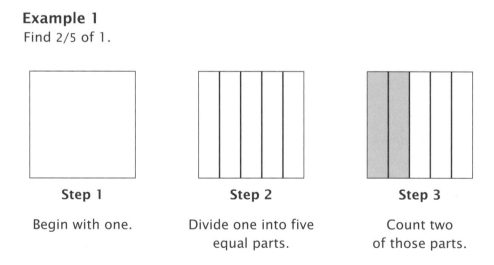

Step 1	Step 2	Step 3
Begin with one.	Divide one into five equal parts.	Count two of those parts.

The bottom number is the denominator, and the top number is the "numberator," or numerator. Stress the three steps: what we start with, how many parts we divide it into, and how many of the divided parts we count or number.

Example 2
Find 3/4 of 1.

 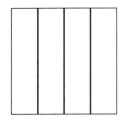 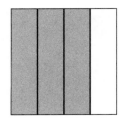

Step 1	**Step 2**	**Step 3**
Begin with one.	Divide one into four equal parts.	Count three of those parts.

Example 3
Find 2/3 of 1.

 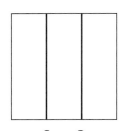 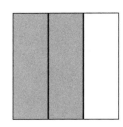

Step 1	**Step 2**	**Step 3**
Begin with one.	Divide one into three equal parts.	Count two of those parts.

Decimals

Math-U-See teaches fractions in depth in *Epsilon*. Decimal numbers are introduced in *Epsilon* and taught in depth in *Zeta*. Students who have used previous levels of Math-U-See should be familiar with using decimals to add and subtract money. If you want to introduce decimals now to prepare students for standardized testing, we suggest that you do so in the context of money.

Students know that there are four quarters in a dollar. They are also familiar with the idea of a fraction as another way to write a division problem. One dollar divided by four is 1/4 of a dollar, or one quarter. Students should know that one quarter is 25 cents, or $0.25 (twenty-five hundredths). Similarly, one dime is 1/10 of a dollar, or 10 cents or $0.10 (ten hundredths). One penny is 1/100 of a dollar, or one cent or $0.01 (one hundredth).

Roman Numerals: D, M, Multiples of 1,000

In this lesson, we will learn two more letter symbols used in the Roman numeral system, as well as the symbol used to write multiples of 1,000. You may want to review the rules and conventions you learned in lesson 28, as they are used for these greater numbers as well.

We are calling the symbol for 1,000 a primary symbol because 1,000 is a power of ten. In other words, we can write 1,000 as $10 \times 10 \times 10$ or as 10×100.

M = 1,000

Using the conventions we learned before, we find that we can use M up to three times in a row, but generally not more than three times. We are calling the symbol for 500 a secondary symbol, as 500 is one half of 1,000. Because D is a secondary symbol, it is customary not to repeat it in a single number.

D = 500

Finally, we can represent 1,000 times a number by writing a line over it.

$\overline{V}$ = 5,000, $\overline{C}$ = 100,000, $\overline{D}$ = 500,000, and $\overline{M}$ = 1,000,000

Here is a chart showing some greater numbers written with Roman numerals.

400	CD	700	DCC
500	D	800	DCCC
600	DC	900	CM

1,000	M	100,000	$\overline{C}$
5,000	$\overline{V}$	700,000	$\overline{DCC}$
10,000	$\overline{X}$	900,000	$\overline{CM}$
50,000	$\overline{L}$		

Study the examples and try writing some Roman numerals for yourself. Look around for examples of Roman numerals in everyday life. Keep in mind the distinction between primary symbols that may be used two or three times in a row and secondary symbols that are usually used only once. These conventions also apply when the line is used over a symbol in order to multiply it by 1,000.

Example 1
Write $\overline{MMMLXX}$CIV using Arabic numerals.

$\overline{MMM}$ is 3,000,000, $\overline{LXX}$ is 70,000, C is 100, and IV is 4, so the answer is 3,070,104.

Example 2
Write $\overline{V}$MMMDCCXCIII using Arabic numerals.

$\overline{V}$ is 5,000, MMM is 3,000, DCC is 700, XC is 90, and III is 3, so the answer is 8,793.

Example 3
Show 243,792 with Roman numerals.

243,792 is 200,000 + 40,000 + 3,000 + 700 + 90 + 2, which is $\overline{CCXL}$MMMDCCXCII.

Example 4
Show 69,508 with Roman numerals.

69,508 is 60,000 + 9,000 + 500 + 8, which is $\overline{LXIX}$DVIII.

Metric Units

The application and enrichment page for lesson 30 (2012 student workbook) has a summary of common metric units. Students should be encouraged to notice metric units used in everyday life and to compare them in a general way to U.S. customary units.

STUDENT SOLUTIONS

Lesson Practice 1A

1. done
2. $3 \times 3 = 9$
3. $6 \times 2 = 12$
   $2 \times 6 = 12$
4. $4 \times 3 = 12$
   $3 \times 4 = 12$
5. $5 \times 5 = 25$
6. $4 \times 2 = 8$
   $2 \times 4 = 8$
7. $6 \times \underline{6} = 36$
8. $2 \times \underline{10} = 20$
9. $4 \times \underline{7} = 28$
10. $5 \times \underline{4} = 20$
11. $7 \times \underline{3} = 21$
12. $8 \times \underline{3} = 24$
13. $7 \times \underline{7} = 49$
14. $5 \times \underline{6} = 30$

Lesson Practice 1B

1. $2 \times 2 = 4$
2. $6 \times 4 = 24$
   $4 \times 6 = 24$
3. $5 \times 2 = 10$
   $2 \times 5 = 10$
4. $5 \times 4 = 20$
   $4 \times 5 = 20$
5. $7 \times 4 = 28$
   $4 \times 7 = 28$
6. $3 \times 2 = 6$
   $2 \times 3 = 6$
7. $3 \times \underline{1} = 3$
8. $2 \times \underline{2} = 4$
9. $1 \times \underline{10} = 10$
10. $8 \times \underline{2} = 16$
11. $7 \times \underline{2} = 14$
12. $5 \times \underline{6} = 30$
13. $10 \times \underline{10} = 100$
14. $5 \times \underline{9} = 45$

Lesson Practice 1C

1. $7 \times 2 = 14$
   $2 \times 7 = 14$
2. $5 \times 3 = 15$
   $3 \times 5 = 15$
3. $7 \times 3 = 21$
   $3 \times 7 = 21$
4. $4 \times 4 = 16$
5. $8 \times 1 = 8$
   $1 \times 8 = 8$
6. $3 \times 4 = 12$
   $4 \times 3 = 12$
7. $8 \times \underline{7} = 56$
8. $9 \times \underline{4} = 36$
9. $6 \times \underline{4} = 24$
10. $3 \times \underline{9} = 27$
11. $5 \times \underline{9} = 45$
12. $6 \times \underline{3} = 18$
13. $10 \times \underline{9} = 90$
14. $2 \times \underline{8} = 16$

Systematic Review 1D

1. $4 \times 2 = 8$
   $2 \times 4 = 8$
2. $5 \times 6 = 30$
   $6 \times 5 = 30$
3. $7 \times \underline{7} = 49$
4. $10 \times \underline{8} = 80$
5. $3 \times 8 = 24$
6. $4 \times \underline{9} = 36$
7. $2 \times 4 = 8$ sq ft
8. $2 \times 2 = 4$ sq ft
9. $6 \times 4 = 24$ sq in
10. $9 \times 3 = 27$ sq in

Systematic Review 1E

1. $6 \times 6 = 36$
2. $4 \times 7 = 28$
   $7 \times 4 = 28$
3. $9 \times \underline{9} = 81$

4. $3 \times \underline{5} = 15$
5. $7 \times \underline{6} = 42$
6. $2 \times \underline{5} = 10$
7. $10 \times 10 = 100$ sq mi
8. $8 \times 4 = 32$ sq in
9. $4 \times 5 = 20$ sq ft
10. $4F = 20;\ F = 5$
11. $3H = 30;\ H = 10$
12. $3 \times 2 = 6$ sq ft

Systematic Review 1F
1. $3 \times 5 = 15$
   $5 \times 3 = 15$
2. $2 \times 6 = 12$
   $6 \times 2 = 12$
3. $5 \times \underline{5} = 25$
4. $7 \times \underline{8} = 56$
5. $9 \times \underline{6} = 54$
6. $3 \times \underline{6} = 18$
7. $6 \times 6 = 36$ sq ft
8. $8 \times 5 = 40$ sq in
9. $7 \times 6 = 42$ sq mi
10. $4D = 24;\ D = 6$
11. $9H = 36;\ H = 4$
12. $6 \times 5 = 30$ sq mi

Lesson Practice 2A
1. $2,4,6,8;\underline{4}$
2. $1,2,3,4,5,6,7,8,9,10;\underline{10}$
3. $2,4,6,8,10,12,14;\underline{7}$
4. $2,4,6;\underline{3}$
5. $1,2,3,4,5,6,7;\underline{7}$
6. $16 \div 2 = \underline{8}$
7. $9 \div 1 = \underline{9}$
8. $4 \div 2 = \underline{2}$
9. $5 \div 1 = \underline{5}$
10. $18 \div 2 = \underline{9}$
11. $8 \div 1 = \underline{8}$
12. $12 \div 2 = \underline{6}$
13. $2 \div 1 = \underline{2}$
14. $4 \div 1 = \underline{4}$

15. $\frac{10}{2} = \underline{5}$
16. $\frac{2}{2} = \underline{1}$
17. $\frac{6}{1} = \underline{6}$
18. $\frac{1}{1} = \underline{1}$
19. $\frac{14}{2} = \underline{7}$
20. $8 \div 2 = \underline{4}$
21. $16 \div 2 = 8$ people
22. $10 \div 2 = 5$ cookies

Lesson Practice 2B
1. $2,4,6,8,10;\underline{5}$
2. $2,4,6,8,10,12,14,16;\underline{8}$
3. $1;\underline{1}$
4. $1,2,3,4,5;\underline{5}$
5. $2,4,6,8,10,12;\underline{6}$
6. $10 \div 2 = \underline{5}$
7. $16 \div 2 = \underline{8}$
8. $14 \div 2 = \underline{7}$
9. $2 \div 2 = 1$
10. $2 \div 1 = \underline{2}$
11. $18 \div 2 = \underline{9}$
12. $9 \div 1 = \underline{9}$
13. $8 \div 2 = \underline{4}$
14. $6 \div 1 = \underline{6}$
15. $\frac{4}{1} = \underline{4}$
16. $\frac{8}{1} = \underline{8}$
17. $\frac{4}{2} = \underline{2}$
18. $\frac{7}{1} = \underline{7}$
19. $\frac{6}{2} = \underline{3}$
20. $14 \div 2 = \underline{7}$
21. $8 \div 1 = 8$ people
22. $8 \div 2 = 4$ people

Lesson Practice 2C

1. 1,2,3,4,5,6,7,8,9;9
2. 2,4,6,8,10,12,14,16,18;9
3. 1,2,3,4;4
4. 2,4,6,8,10,12,14,16,18,20;10
5. 2,4;2
6. $2 \div 2 = 1$
7. $8 \div 2 = 4$
8. $6 \div 2 = 3$
9. $4 \div 2 = 2$
10. $6 \div 1 = 6$
11. $12 \div 2 = 6$
12. $10 \div 1 = 10$
13. $16 \div 2 = 8$
14. $7 \div 1 = 7$
15. $\frac{14}{2} = 7$
16. $\frac{1}{1} = 1$
17. $\frac{18}{2} = 9$
18. $\frac{9}{1} = 9$
19. $\frac{20}{2} = 10$
20. $10 \div 2 = 5$
21. $12 \div 2 = 6$ people
22. $6 \div 1 = 6$ people

Systematic Review 2D

1. $4 \div 2 = 2$
2. $8 \div 1 = 8$
3. $4 \div 1 = 4$
4. $18 \div 2 = 9$
5. $2 \div 2 = 1$
6. $2 \div 1 = 2$
7. $\frac{14}{2} = 7$
8. $\frac{16}{2} = 8$
9. $\frac{5}{1} = 5$
10. $\begin{array}{r} 5 \\ \times\ 6 \\ \hline 30 \end{array}$

11. $\begin{array}{r} 10 \\ \times\ 9 \\ \hline 90 \end{array}$

12. $10 \times 6 = 60$
13. $(5)(7) = 35$
14. $5 \times 7 = 35$
15. $5 \times 9 = 45$
16. $10 \times 1 = 10$
17. $10 \times 4 = 40$
18. $5 \times 5 = 25$ sq ft
19. $5 \times 3 = 15$ sq mi
20. $3 \times 10 = 30$ sq in
21. $18 \div 2 = 9$
22. $18 \div 2 = 9$ piles

Systematic Review 2E

1. $6 \div 1 = 6$
2. $8 \div 2 = 4$
3. $9 \div 1 = 9$
4. $18 \div 2 = 9$
5. $2 \div 2 = 1$
6. $2 \div 1 = 2$
7. $\frac{12}{1} = 12$
8. $\frac{10}{2} = 5$
9. $\frac{1}{1} = 1$
10. $5 \times 1 = 5$
11. $8 \times 10 = 80$
12. $7 \times 5 = 35$
13. $3 \times 10 = 30$
14. $\begin{array}{r} 8 \\ \times\ 5 \\ \hline 40 \end{array}$

15. $\begin{array}{r} 10 \\ \times\ 7 \\ \hline 70 \end{array}$

16. $2 \times 10 = 20$
17. $10 \cdot 10 = 100$
18. $2 \times 2 = 4$ sq ft
19. $5 \times 10 = 50$ sq mi
20. $3 \times 5 = 15$ sq in

21. $8 \div 1 = 8$
22. $20 \div 2 = 10$ cages

Systematic Review 2F

1. $5 \div 1 = \underline{5}$
2. $16 \div 2 = \underline{8}$
3. $4 \div 2 = \underline{2}$
4. $8 \div 1 = \underline{8}$
5. $14 \div 2 = \underline{7}$
6. $12 \div 2 = \underline{6}$
7. $\dfrac{18}{2} = \underline{9}$
8. $\dfrac{10}{2} = \underline{5}$
9. $\dfrac{20}{2} = \underline{10}$
10. $5 \times \underline{4} = 20$
11. $5 \times \underline{10} = 50$
12. $9 \times \underline{10} = 90$
13. $5 \times \underline{8} = 40$
14.
$$\begin{array}{r} 9 \\ \times\ 5 \\ \hline 45 \end{array}$$
15.
$$\begin{array}{r} 10 \\ \times\ 4 \\ \hline 40 \end{array}$$
16. $2 \cdot 5 = \underline{10}$
17. $(10)(8) = \underline{80}$
18. $10 \times 10 = 100$ sq in
19. $1 \times 1 = 1$ sq mi
20. $6 \times 5 = 30$ sq ft
21. $2 \div 2 = 1$
22. $14 \div 2 = 7$ stickers

Lesson Practice 3A

1. 10,20,30,40,50,60,70,80;$\underline{8}$
2. 10,20,30,40,50,60,70,80,90,100;$\underline{10}$
3. 10,20,30,40,50,60;$\underline{6}$
4. 10;$\underline{1}$
5. $50 \div 10 = \underline{5}$
6. $100 \div 10 = \underline{10}$
7. $30 \div 10 = \underline{3}$

8. $90 \div 10 = \underline{9}$
9. $20 \div 10 = \underline{2}$
10. $40 \div 10 = \underline{4}$
11. $10 \div 10 = \underline{1}$
12. $70 \div 10 = \underline{7}$
13. $50 \div 10 = \underline{5}$
14. $\dfrac{30}{10} = \underline{3}$
15. $\dfrac{90}{10} = \underline{9}$
16. $60 \div 10 = \underline{6}$
17. $80 \div 10 = 8$ teams
18. $\$40 \div \$10 = 4$ books

Lesson Practice 3B

1. 10,20;$\underline{2}$
2. 10,20,30,40;$\underline{4}$
3. 10,20,30,40,50,60,70,80,90;$\underline{9}$
4. 10,20,30;$\underline{3}$
5. $60 \div 10 = \underline{6}$
6. $90 \div 10 = \underline{9}$
7. $50 \div 10 = \underline{5}$
8. $30 \div 10 = \underline{3}$
9. $70 \div 10 = \underline{7}$
10. $10 \div 10 = \underline{1}$
11. $40 \div 10 = \underline{4}$
12. $20 \div 10 = \underline{2}$
13. $90 \div 10 = \underline{9}$
14. $\dfrac{100}{10} = \underline{10}$
15. $\dfrac{50}{10} = \underline{5}$
16. $10 \div 10 = \underline{1}$
17. $70 \div 10 = 7$ jelly beans
18. $30 \div 10 = 3$ sheets

Lesson Practice 3C

1. 10,20,30,40,50;$\underline{5}$
2. 10,20,30,40,50,60,70;$\underline{7}$
3. 10,20,30,40,50,60,70,80,90,100;$\underline{10}$
4. 10,20,30,40,50,60;$\underline{6}$
5. $90 \div 10 = \underline{9}$

6. $40 \div 10 = \underline{4}$

7. $100 \div 10 = \underline{10}$

8. $50 \div 10 = \underline{5}$

9. $80 \div 10 = \underline{8}$

10. $60 \div 10 = \underline{6}$

11. $20 \div 10 = \underline{2}$

12. $30 \div 10 = \underline{3}$

13. $10 \div 10 = \underline{1}$

14. $\dfrac{70}{10} = \underline{7}$

15. $\dfrac{40}{10} = \underline{4}$

16. $30 \div 10 = \underline{3}$

17. $50 \div 10 = 5$ hours

18. $100 \div 10 = 10$ pennies

Systematic Review 3D

1. $10 \div 10 = \underline{1}$

2. $30 \div 10 = \underline{3}$

3. $60 \div 10 = \underline{6}$

4. $70 \div 10 = \underline{7}$

5. $5 \div 1 = \underline{5}$

6. $14 \div 2 = \underline{7}$

7. $100 \div 10 = \underline{10}$

8. $10 \div 1 = \underline{10}$

9. $18 \div 2 = \underline{9}$

10. $8 \div 1 = \underline{8}$

11. $\dfrac{50}{10} = \underline{5}$

12. $\dfrac{16}{2} = \underline{8}$

13. $\begin{array}{r} 9 \\ \times\ 3 \\ \hline 27 \end{array}$

14. $\begin{array}{r} 3 \\ \times\ 4 \\ \hline 12 \end{array}$

15. $3 \times 7 = \underline{21}$

16. $10 \cdot 3 = \underline{30}$

17. $12 \div 2 = 6$ qt

18. $8 \times 3 = 24$ ft

19. $5 \times 8 = 40$ sq in

20. $12 \div 2 = 6$ boxes

Systematic Review 3E

1. $20 \div 10 = \underline{2}$

2. $40 \div 10 = \underline{4}$

3. $90 \div 10 = \underline{9}$

4. $80 \div 10 = \underline{8}$

5. $7 \div 1 = \underline{7}$

6. $20 \div 2 = \underline{10}$

7. $10 \div 10 = \underline{1}$

8. $6 \div 1 = \underline{6}$

9. $8 \div 2 = \underline{4}$

10. $4 \div 1 = \underline{4}$

11. $\dfrac{6}{2} = \underline{3}$

12. $\dfrac{14}{2} = \underline{7}$

13. $\begin{array}{r} 3 \\ \times 3 \\ \hline 9 \end{array}$

14. $\begin{array}{r} 5 \\ \times\ 3 \\ \hline 15 \end{array}$

15. $(2)(3) = \underline{6}$

16. $6 \times 3 = \underline{18}$

17. $14 \div 2 = 7$ qt

18. $5 \times 3 = 15$ ft

19. $1 \times 2 = 2$ sq mi

20. $18 \div 2 = 9$ people

Systematic Review 3F

1. $70 \div 10 = \underline{7}$

2. $60 \div 10 = \underline{6}$

3. $50 \div 10 = \underline{5}$

4. $30 \div 10 = \underline{3}$

5. $9 \div 1 = \underline{9}$

6. $16 \div 2 = \underline{8}$

7. $4 \div 2 = \underline{2}$

8. $3 \div 1 = \underline{3}$

9. $10 \div 2 = \underline{5}$

10. $12 \div 2 = \underline{6}$

11. $\dfrac{2}{2} = \underline{1}$

12. $\dfrac{18}{2} = \underline{9}$

13.
$$\begin{array}{r} 4 \\ \times\ 3 \\ \hline 12 \end{array}$$

14.
$$\begin{array}{r} 3 \\ \times\ 8 \\ \hline 24 \end{array}$$

15. $7 \cdot 3 = \underline{21}$
16. $3 \times 9 = \underline{27}$
17. $16 \div 2 = 8$ qt
18. $6 \times 3 = 18$ sq yd
19. $6 \times 3 = 18$ ft long
    $3 \times 3 = 9$ ft wide
20. $3 \times 10 = 30$ cookies
    $30 \div 10 = 3$ cookies

Lesson Practice 4A

1. 5,10,15,20,25,30,35;$\underline{7}$
2. 5,10,15,20,25,30,35,40,45,50;$\underline{10}$
3. 5;$\underline{1}$
4. 3,6,9,12,15,18;$\underline{6}$
5. $45 \div 5 = \underline{9}$
6. $20 \div 5 = \underline{4}$
7. $15 \div 5 = \underline{3}$
8. $30 \div 3 = \underline{10}$
9. $12 \div 3 = \underline{4}$
10. $27 \div 3 = \underline{9}$
11. $30 \div 5 = \underline{6}$
12. $10 \div 5 = \underline{2}$
13. $40 \div 5 = \underline{8}$
14. $\frac{25}{5} = \underline{5}$
15. $\frac{24}{3} = \underline{8}$
16. $\frac{21}{3} = \underline{7}$
17. $9 \div 3 = 3$ pieces
18. $15 \div 3 = 5$ cages

Lesson Practice 4B

1. 5,10,15,20,25,30,35,40;$\underline{8}$
2. 3;$\underline{1}$
3. 5,10,15,20,25;$\underline{5}$

4. 3,6,9,12,15,18,21,24,27;$\underline{9}$
5. $5 \div 5 = \underline{1}$
6. $35 \div 5 = \underline{7}$
7. $50 \div 5 = \underline{10}$
8. $6 \div 3 = \underline{2}$
9. $24 \div 3 = \underline{8}$
10. $18 \div 3 = \underline{6}$
11. $20 \div 5 = \underline{4}$
12. $45 \div 5 = \underline{9}$
13. $21 \div 3 = \underline{7}$
14. $\frac{15}{3} = \underline{5}$
15. $\frac{30}{3} = \underline{10}$
16. $\frac{30}{5} = \underline{6}$
17. $15 \div 5 = 3$ five-dollar bills
18. $21 \div 3 = 7$ apples

Lesson Practice 4C

1. 3,6,9;$\underline{3}$
2. 3,6,9,12,15,18,21,24;$\underline{8}$
3. 5,10;$\underline{2}$
4. 5,10,15,20;$\underline{4}$
5. $12 \div 3 = \underline{4}$
6. $27 \div 3 = \underline{9}$
7. $40 \div 5 = \underline{8}$
8. $15 \div 3 = \underline{5}$
9. $15 \div 5 = \underline{3}$
10. $30 \div 3 = \underline{10}$
11. $25 \div 5 = \underline{5}$
12. $50 \div 5 = \underline{10}$
13. $18 \div 3 = \underline{6}$
14. $\frac{30}{5} = \underline{6}$
15. $\frac{21}{3} = \underline{7}$
16. $\frac{5}{5} = \underline{1}$
17. $24 \div 3 = 8$ feet
18. $35 \div 5 = 7$ hands

Systematic Review 4D

1. $6 \div 3 = \underline{2}$
2. $30 \div 5 = \underline{6}$
3. $18 \div 3 = \underline{6}$
4. $45 \div 5 = \underline{9}$
5. $15 \div 3 = \underline{5}$
6. $25 \div 5 = \underline{5}$
7. $40 \div 10 = \underline{4}$
8. $16 \div 2 = \underline{8}$
9. $24 \div 3 = \underline{8}$
10. $30 \div 3 = \underline{10}$
11. $\frac{14}{2} = \underline{7}$
12. $\frac{40}{5} = \underline{8}$
13. $4 \times \underline{3} = 12$
14. $6 \times \underline{5} = 30$
15. $5 \times \underline{3} = 15$
16. $8 \times \underline{3} = 24$
17. $\begin{array}{r} 25 \\ +34 \\ \hline 59 \end{array}$
18. $\begin{array}{r} {\scriptstyle 1} \\ 78 \\ +34 \\ \hline 112 \end{array}$
19. $\begin{array}{r} {\scriptstyle 1} \\ 49 \\ +51 \\ \hline 100 \end{array}$
20. $\begin{array}{r} {\scriptstyle 1} \\ 65 \\ +15 \\ \hline 80 \end{array}$
21. $\$39 + \$28 = \$67$
22. $\$50 \div \$5 = 10$ days

Systematic Review 4E

1. $12 \div 3 = \underline{4}$
2. $35 \div 5 = \underline{7}$
3. $15 \div 5 = \underline{3}$
4. $9 \div 3 = \underline{3}$
5. $21 \div 3 = \underline{7}$
6. $10 \div 2 = \underline{5}$

7. $27 \div 3 = \underline{9}$
8. $60 \div 10 = \underline{6}$
9. $8 \div 2 = \underline{4}$
10. $3 \div 1 = \underline{3}$
11. $\frac{50}{5} = \underline{10}$
12. $\frac{3}{3} = \underline{1}$
13. $5 \times \underline{7} = 35$
14. $3 \times \underline{7} = 21$
15. $7 \times \underline{10} = 70$
16. $6 \times \underline{3} = 18$
17. $\begin{array}{r} {\scriptstyle 1} \\ 13 \\ +19 \\ \hline 32 \end{array}$
18. $\begin{array}{r} {\scriptstyle 1} \\ 28 \\ +49 \\ \hline 77 \end{array}$
19. $\begin{array}{r} 26 \\ +72 \\ \hline 98 \end{array}$
20. $\begin{array}{r} {\scriptstyle 1} \\ 47 \\ +38 \\ \hline 85 \end{array}$
21. $20 \div 5 = 4$ stickers
22. $27 \div 3 = 9$ pages
23. $10 \times 2 = 20$ pints
    $20 + 3 = 23$ pints
24. $7 \times 2 = 14$ notes
    $14 + 11 = 25$ people thanked

Systematic Review 4F

1. $10 \div 5 = \underline{2}$
2. $24 \div 3 = \underline{8}$
3. $30 \div 3 = \underline{10}$
4. $40 \div 5 = \underline{8}$
5. $35 \div 5 = \underline{7}$
6. $15 \div 5 = \underline{3}$
7. $12 \div 3 = \underline{4}$
8. $18 \div 3 = \underline{6}$
9. $18 \div 2 = \underline{9}$

10. $70 \div 10 = \underline{7}$

11. $\frac{27}{3} = \underline{9}$

12. $\frac{5}{1} = \underline{5}$

13. $5 \times \underline{10} = 50$

14. $3 \times \underline{3} = 9$

15. $10 \times \underline{9} = 90$

16. $9 \times \underline{1} = 9$

17.
```
  81
+18
  99
```

18.
```
 1
 37
+37
 74
```

19.
```
 1
 42
+29
 71
```

20.
```
 74
+25
 99
```

21. $9 \times 5 = 45$ books

22. $3 \times \$5 = \$15$
$\$15 \div 5 = \$3$ per person

23. $18 \div 3 = 6$ pieces

24. $46 + 28 = 74$ mi

Lesson Practice 5A

1. true
2. true
3. false
4. false

The answers to 5 and 6 may be drawn in any position.

5. $\perp$
6. $\parallel$
7. no
8. no
9. yes
10. 2
11. perpendicular
12. parallel

Lesson Practice 5B

1. false
2. false
3. true
4. true
5. $\parallel$
6. $\perp$
7. perpendicular
8. no
9. yes
10. no
11. parallel
12. perpendicular

Lesson Practice 5C

1. true
2. true
3. false
4. false
5. $\perp$
6. $\parallel$
7. no
8. no
9. yes
10. 4
11. perpendicular
12. parallel

Systematic Review 5D

1.
```
 45
+62
107
```

2.
```
 1
 17
+34
 51
```

3.
```
 1
 55
+55
110
```

4.
```
 1
 29
+71
100
```

5. $9 \times \underline{4} = 36$
6. $9 \times \underline{6} = 54$
7. $9 \times \underline{9} = 81$
8. $9 \times \underline{7} = 63$
9. $60 \div 10 = \underline{6}$
10. $15 \div 3 = \underline{5}$
11. $\frac{14}{2} = \underline{7}$
12. $\frac{30}{10} = \underline{3}$
13. $6 \div 3 = \underline{2}$
14. $45 \div 5 = \underline{9}$
15. $18 \div 2 = \underline{9}$
16. $35 \div 5 = \underline{7}$
17. yes, in most cases
18. yes, in most cases
19. $20 \div 5 = 4$ bags
20. $25 + 38 = 63$ problems

Systematic Review 5E
1.
$$\begin{array}{r} \overset{1}{19} \\ +88 \\ \hline 107 \end{array}$$
2.
$$\begin{array}{r} 54 \\ +13 \\ \hline 67 \end{array}$$
3.
$$\begin{array}{r} \overset{1}{43} \\ +67 \\ \hline 110 \end{array}$$
4.
$$\begin{array}{r} \overset{1}{74} \\ +48 \\ \hline 122 \end{array}$$
5. $9 \times \underline{8} = 72$
6. $9 \times \underline{10} = 90$
7. $9 \times \underline{2} = 18$
8. $9 \times \underline{10} = 90$
9. $50 \div 10 = \underline{5}$
10. $15 \div 5 = \underline{3}$
11. $\frac{18}{3} = \underline{6}$
12. $\frac{16}{2} = \underline{8}$

13. $30 \div 10 = \underline{3}$
14. $25 \div 5 = \underline{5}$
15. $27 \div 3 = \underline{9}$
16. $12 \div 2 = \underline{6}$
17. $\|; \perp$
18. no
19. $20 \div 2 = 10$ gumdrops
20. $9 \times 9 = 81$ sq ft

Systematic Review 5F
1.
$$\begin{array}{r} 24 \\ +35 \\ \hline 59 \end{array}$$
2.
$$\begin{array}{r} \overset{1}{13} \\ +19 \\ \hline 32 \end{array}$$
3.
$$\begin{array}{r} \overset{1}{81} \\ +79 \\ \hline 160 \end{array}$$
4.
$$\begin{array}{r} 65 \\ +42 \\ \hline 107 \end{array}$$
5. $9 \times \underline{2} = 18$
6. $9 \times \underline{6} = 54$
7. $9 \times \underline{8} = 72$
8. $9 \times \underline{1} = 9$
9. $40 \div 5 = \underline{8}$
10. $70 \div 10 = \underline{7}$
11. $\frac{21}{3} = \underline{7}$
12. $\frac{30}{3} = \underline{10}$
13. $100 \div 10 = \underline{10}$
14. $20 \div 5 = \underline{4}$
15. $24 \div 3 = \underline{8}$
16. $8 \div 2 = \underline{4}$
17. parallel, $\|$; perpendicular, $\perp$
18. no
19. $27 \div 3 = 9$ yd
20. $16 + 14 = 30$ children

Lesson Practice 6A

1. 9,18,27,36;$\underline{4}$
2. 9,18,27;$\underline{3}$
3. 9,18;$\underline{2}$
4. 9,18,27,36,45,54;$\underline{6}$
5. $81 \div 9 = \underline{9}$
6. $18 \div 9 = \underline{2}$
7. $63 \div 9 = \underline{7}$
8. $45 \div 9 = \underline{5}$
9. $90 \div 9 = \underline{10}$
10. $27 \div 9 = \underline{3}$
11. $9 \div 9 = \underline{1}$
12. $54 \div 9 = \underline{6}$
13. $72 \div 9 = \underline{8}$
14. $\frac{36}{9} = \underline{4}$
15. $\frac{81}{9} = \underline{9}$
16. $\frac{45}{9} = \underline{5}$
17. $\$90 \div 9 = \$10$ per day
18. $36 \div 9 = 4$ pieces

Lesson Practice 6B

1. 9,18,27,36,45;$\underline{5}$
2. 9,18;$\underline{2}$
3. 9,18,27;$\underline{3}$
4. 9,18,27,36,45,54,63,72;$\underline{8}$
5. $27 \div 9 = \underline{3}$
6. $54 \div 9 = \underline{6}$
7. $36 \div 9 = \underline{4}$
8. $72 \div 9 = \underline{8}$
9. $45 \div 9 = \underline{5}$
10. $9 \div 9 = \underline{1}$
11. $18 \div 9 = \underline{2}$
12. $81 \div 9 = \underline{9}$
13. $63 \div 9 = \underline{7}$
14. $\frac{90}{9} = \underline{10}$
15. $\frac{54}{9} = \underline{6}$
16. $\frac{36}{9} = \underline{4}$

17. $81 \div 9 = 9$ stories
18. $27 \div 9 = 3$ cakes

Lesson Practice 6C

1. 9,18,27,36,45,54,63,72,81,90;$\underline{10}$
2. 9,18,27,36,45,54,63,72,81;$\underline{9}$
3. 9;$\underline{1}$
4. 9,18,27,36,45,54,63;$\underline{7}$
5. $54 \div 9 = \underline{6}$
6. $9 \div 9 = \underline{1}$
7. $72 \div 9 = \underline{8}$
8. $45 \div 9 = \underline{5}$
9. $27 \div 9 = \underline{3}$
10. $36 \div 9 = \underline{4}$
11. $18 \div 9 = \underline{2}$
12. $63 \div 9 = \underline{7}$
13. $45 \div 9 = \underline{5}$
14. $\frac{72}{9} = \underline{8}$
15. $\frac{27}{9} = \underline{3}$
16. $\frac{81}{9} = \underline{9}$
17. $18 \div 2 = 9$ years
18. $54 \div 9 = 6$ teams

Systematic Review 6D

1. $54 \div 9 = \underline{6}$
2. $45 \div 9 = \underline{5}$
3. $63 \div 9 = \underline{7}$
4. $24 \div 3 = \underline{8}$
5. $35 \div 5 = \underline{7}$
6. $60 \div 10 = \underline{6}$
7. $36 \div 9 = \underline{4}$
8. $18 \div 2 = \underline{9}$
9. $72 \div 9 = \underline{8}$
10. $90 \div 9 = \underline{10}$
11. $\frac{25}{5} = \underline{5}$
12. $\frac{21}{3} = \underline{7}$
13. $5 \times 1 = 5$
14. $9 \times \underline{9} = 81$

15. $3 \times \underline{4} = 12$

16. $10 \times \underline{10} = 100$

17. done

18.
$$\begin{array}{r} {}^{6}\!\!\not7\,{}^{1}4 \\ -\ 3\ 8 \\ \hline 3\ 6 \end{array}$$

19.
$$\begin{array}{r} 5\ 9 \\ -4\ 1 \\ \hline 1\ 8 \end{array}$$

20.
$$\begin{array}{r} 6\ 7 \\ -2\ 5 \\ \hline 4\ 2 \end{array}$$

21. $81 \div 9 = 9$ wreaths

22. $95¢ - 87¢ = 8¢$ change

Systematic Review 6E

1. $36 \div 9 = \underline{4}$

2. $54 \div 9 = \underline{6}$

3. $18 \div 9 = \underline{2}$

4. $9 \div 9 = \underline{1}$

5. $15 \div 3 = \underline{5}$

6. $20 \div 5 = \underline{4}$

7. $40 \div 10 = \underline{4}$

8. $12 \div 2 = \underline{6}$

9. $27 \div 9 = \underline{3}$

10. $81 \div 9 = \underline{9}$

11. $\dfrac{72}{9} = \underline{8}$

12. $\dfrac{12}{3} = \underline{4}$

13. $2 \times \underline{9} = 18$

14. $10 \times \underline{5} = 50$

15. $3 \times \underline{7} = 21$

16. $5 \times \underline{9} = 45$

17.
$$\begin{array}{r} {}^{3}\!\!\not4\,{}^{1}3 \\ -\ 1\ 9 \\ \hline 2\ 4 \end{array}$$

18.
$$\begin{array}{r} {}^{6}\!\!\not7\,{}^{1}8 \\ -\ 5\ 9 \\ \hline 1\ 9 \end{array}$$

19.
$$\begin{array}{r} {}^{1} \\ 2\ 6 \\ +7\ 5 \\ \hline 1\ 0\ 1 \end{array}$$

20.
$$\begin{array}{r} {}^{5}\!\!\not6\,{}^{1}7 \\ -\ 3\ 8 \\ \hline 2\ 9 \end{array}$$

21. $25 + 25 = 50$ pints

$50 \div 10 = 5$ pints each

22. $14 \div 2 = 7$ qt

23. $27 \div 3 = 9$ yd

24. $32 - 14 = 18$ oranges

Systematic Review 6F

1. $90 \div 9 = \underline{10}$

2. $18 \div 9 = \underline{2}$

3. $36 \div 9 = \underline{4}$

4. $54 \div 9 = \underline{6}$

5. $27 \div 3 = \underline{9}$

6. $40 \div 5 = \underline{8}$

7. $72 \div 9 = \underline{8}$

8. $45 \div 5 = \underline{9}$

9. $80 \div 10 = \underline{8}$

10. $16 \div 2 = \underline{8}$

11. $\dfrac{63}{9} = \underline{7}$

12. $\dfrac{45}{9} = \underline{5}$

13. $9 \times \underline{8} = 72$

14. $3 \times \underline{8} = 24$

15. $5 \times \underline{10} = 50$

16. $9 \times \underline{0} = 0$

17.
$$\begin{array}{r} {}^{1} \\ 8\ 2 \\ +1\ 8 \\ \hline 1\ 0\ 0 \end{array}$$

18.
$$\begin{array}{r} {}^{2}\!\!\not3\,{}^{1}7 \\ -\ 2\ 8 \\ \hline 9 \end{array}$$

19.
$$\begin{array}{r} {}^{5}\!\!\not6\,{}^{1}6 \\ -\ 3\ 9 \\ \hline 2\ 7 \end{array}$$

20.

$$\begin{array}{r} 75 \\ +24 \\ \hline 99 \end{array}$$

21. $8 + 4 = 12$ pt

$12 \div 2 = 6$ qt

22. $9 \times 3 = 27$ ft

23. $45 \div 9 = 5$ vans

24. $28¢ + 35¢ = 63¢$

$63¢ \div 9¢ = 7$ treats

Lesson Practice 7A

1. done
2. $4 \times 6 = 24$ sq ft
3. $1 \times 2 = 2$ sq ft
4. $5 \times 5 = 25$ sq in
5. $3 \times 7 = 21$ sq in
6. $4 \times 10 = 40$ sq ft
7. $7 \times 5 = 35$ sq in
8. $9 \times 5 = 45$ sq yd

Lesson Practice 7B

1. $5 \times 10 = 50$ sq ft
2. $3 \times 9 = 27$ sq in
3. $9 \times 4 = 36$ sq in
4. $6 \times 10 = 60$ sq ft
5. $6 \times 7 = 42$ sq ft
6. $2 \times 10 = 20$ sq in
7. $10 \times 10 = 100$ sq in
8. $8 \times 5 = 40$ sq in

Lesson Practice 7C

1. $9 \times 7 = 63$ sq ft
2. $4 \times 5 = 20$ sq in
3. $10 \times 8 = 80$ sq in
4. $5 \times 6 = 30$ sq ft
5. $4 \times 7 = 28$ sq ft
6. $9 \times 5 = 45$ sq in
7. $8 \times 9 = 72$ sq yd
8. $10 \times 9 = 90$ sq yd

Systematic Review 7D

1. $9 \times 6 = 54$ sq ft
2. $6 \times 8 = 48$ sq in
3. $10 \times 5 = 50$ sq in
4. $6 \times 6 = 36$ sq mi
5. $36 \div 9 = \underline{4}$
6. $45 \div 5 = \underline{9}$
7. $18 \div 2 = \underline{9}$
8. $63 \div 9 = \underline{7}$
9. $27 \div 3 = \underline{9}$
10. $35 \div 5 = \underline{7}$
11. $\dfrac{18}{3} = \underline{6}$
12. $\dfrac{50}{10} = \underline{5}$
13. $6 \times \underline{7} = 42$
14. $6 \times \underline{5} = 30$
15. $4 \times \underline{8} = 32$
16. $6 \times \underline{2} = 12$
17.

$$\begin{array}{r} {}^{1} \\ 38 \\ +26 \\ \hline 64 \end{array}$$

18.

$$\begin{array}{r} 15 \\ +84 \\ \hline 99 \end{array}$$

19.

$$\begin{array}{r} {}^{3}\cancel{4}\,{}^{1}5 \\ -\ 1\ 6 \\ \hline 2\ 9 \end{array}$$

20.

$$\begin{array}{r} {}^{6}\cancel{7}\,{}^{1}1 \\ -\ 5\ 6 \\ \hline 1\ 5 \end{array}$$

21. $6 \times 5 = 30$ sq in

$4 \times 4 = 16$ sq in

$30 > 16$

parallelogram

22. $43 - 28 = 15$ barrettes

Systematic Review 7E

1. $3 \times 2 = 6$ sq in
2. $7 \times 9 = 63$ sq ft
3. $8 \times 4 = 32$ sq mi
4. $10 \times 10 = 100$ sq ft

5. $12 \div 3 = \underline{4}$
6. $54 \div 9 = \underline{6}$
7. $24 \div 3 = \underline{8}$
8. $25 \div 5 = \underline{5}$
9. $14 \div 2 = \underline{7}$
10. $8 \div 1 = \underline{8}$
11. $\frac{81}{9} = \underline{9}$
12. $\frac{21}{3} = \underline{7}$
13. $4 \times \underline{6} = 24$
14. $6 \times \underline{10} = 60$
15. $6 \times \underline{7} = 42$
16. $4 \times \underline{7} = 28$
17.
```
  71
+62
 133
```
18.
```
 ³4̶ ¹3
 - 2 5
    1 8
```
19.
```
  92
+11
 103
```
20.
```
  ¹
  57
+46
 103
```
21. parallel
22. $5 \times 3 = 15$ sq yd

Systematic Review 7F

1. $6 \times 7 = 42$ sq ft
2. $3 \times 8 = 24$ sq in
3. $10 \times 9 = 90$ sq ft
4. $3 \times 3 = 9$ sq mi
5. $27 \div 9 = \underline{3}$
6. $15 \div 3 = \underline{5}$
7. $30 \div 5 = \underline{6}$
8. $16 \div 2 = \underline{8}$
9. $72 \div 9 = \underline{8}$
10. $90 \div 10 = \underline{9}$
11. $\frac{20}{2} = \underline{10}$

12. $\frac{45}{9} = \underline{5}$
13. $4 \times \underline{8} = 32$
14. $6 \times \underline{8} = 48$
15. $6 \times \underline{6} = 36$
16. $4 \times \underline{4} = 16$
17.
```
 ¹2̶¹
 - 9
  1 2
```
18.
```
  ¹
  76
+54
 130
```
19.
```
  33
+45
  78
```
20.
```
 ⁵6̶ ¹4
 - 2 5
    3 9
```
21. $14 \div 2 = 7$ qt
22. $30 - 16 = 14$ books

Lesson Practice 8A

1. 6,12,18;$\underline{3}$
2. 6,12,18,24,30,36,42,48,54;$\underline{9}$
3. 6,12;$\underline{2}$
4. 6,12,18,24,30,36,42,48,54,60;$\underline{10}$
5. $12 \div 6 = \underline{2}$
6. $6 \div 6 = \underline{1}$
7. $24 \div 6 = \underline{4}$
8. $36 \div 6 = \underline{6}$
9. $42 \div 6 = \underline{7}$
10. $18 \div 6 = \underline{3}$
11. $60 \div 6 = \underline{10}$
12. $24 \div 6 = \underline{4}$
13. $42 \div 6 = \underline{7}$
14. $\frac{54}{6} = \underline{9}$
15. $\frac{30}{6} = \underline{5}$
16. $48 \div \underline{6} = 8$

17. $24 \div 6 = 4$ ants

18. $\$30 \div 6 = \$5$ a day

Lesson Practice 8B

1. 6,12,18,24,30;5

2. 6;1

3. 6,12,18,24;4

4. 6,12,18,24,30,36,42,48;8

5. $36 \div 6 = 6$

6. $60 \div 6 = 10$

7. $30 \div 6 = 5$

8. $18 \div 6 = 3$

9. $54 \div 6 = 9$

10. $42 \div 6 = 7$

11. $6 \div 6 = 1$

12. $24 \div 6 = 4$

13. $18 \div 6 = 3$

14. $\frac{30}{6} = 5$

15. $\frac{48}{6} = 8$

16. $12 \div 6 = 2$

17. $60 \div 6 = 10$ songs

18. $\$54 \div 6 = \$9$ each hour

Lesson Practice 8C

1. 6,12,18,24,30,36,42,48,54;9

2. 6,12,18,24,30,36;6

3. 6,12,18,24,30,36,42,48,54,60;10

4. 6,12,18,24,30,36,42;7

5. $18 \div 6 = 3$

6. $54 \div 6 = 9$

7. $6 \div 6 = 1$

8. $30 \div 6 = 5$

9. $12 \div 6 = 2$

10. $24 \div 6 = 4$

11. $42 \div 6 = 7$

12. $36 \div 6 = 6$

13. $48 \div 6 = 8$

14. $\frac{60}{6} = 10$

15. $\frac{54}{6} = 9$

16. $\frac{12}{6} = 2$

17. $\$48 \div 6 = \$8$ per friend

18. $18 \div 6 = 3$ ft

$3 \div 3 = 1$ yd

Systematic Review 8D

1. $18 \div 6 = 3$

2. $42 \div 6 = 7$

3. $54 \div 6 = 9$

4. $24 \div 3 = 8$

5. $25 \div 5 = 5$

6. $18 \div 2 = 9$

7. $54 \div 9 = 6$

8. $60 \div 10 = 6$

9. $48 \div 6 = 8$

10. $72 \div 9 = 8$

11. $\frac{21}{3} = 7$

12. $\frac{35}{5} = 7$

13. $12 \times 6 = 72$ sq ft

14. $7 \times 3 = 21$ sq in

15. $4 \times 4 = 16$ sq in

16.
```
   2 3        20 + 3
 × 3 6      ×  30 + 6
 ─────      ─────────
  1 1       100   10
  1 2 8     100 + 20 + 8
  6 9       600 + 90 +
 ─────      ─────────────
  8 2 8     800 + 20 + 8
```

17.
```
    7 8              70 + 8
  × 3 4            × 30 + 4
  ─────           ───────────
    1          100 200    30
   2 3            200  + 80 + 2
   2 8 2      2000 + 100  + 40
   2 1 4      ─────────────────
  ───────    2000 + 600 + 50 + 2
  2 6 5 2
```

18.
```
    6 5            60 + 5
  × 1 5          × 10 + 5
  ─────          ─────────
    2              20
  3 0 5      300 + 00 + 5
    6 5      600 + 50 +
  ───────    ────────────
  9 7 5      900 + 70 + 5
```

19. $12 \times 15 = 180$ baby mice
20. $61 - 45 = 16$ sq ft
21. $36 \div 6 = 6$ afghans
22. $\$39 + \$28 = \$67$

Systematic Review 8E

1. $12 \div 6 = \underline{2}$
2. $60 \div 6 = \underline{10}$
3. $42 \div 6 = \underline{7}$
4. $24 \div 6 = \underline{4}$
5. $27 \div 9 = \underline{3}$
6. $40 \div 5 = \underline{8}$
7. $20 \div 10 = \underline{2}$
8. $12 \div 3 = \underline{4}$
9. $15 \div 3 = \underline{5}$
10. $30 \div 6 = \underline{5}$
11. $\dfrac{6}{6} = \underline{1}$
12. $\dfrac{12}{2} = \underline{6}$

13.
$$\begin{array}{r} {}^113 \\ +19 \\ \hline 32 \end{array}$$

14.
$$\begin{array}{r} {}^128 \\ +49 \\ \hline 77 \end{array}$$

15.
$$\begin{array}{r} {}^6\!\!\!\not{7}\,{}^1 2 \\ -\,2\,6 \\ \hline 4\,6 \end{array}$$

16.
$$\begin{array}{r} {}^3\!\!\!\not{4}\,{}^1 7 \\ -\,3\,8 \\ \hline 9 \end{array}$$

17.
$$\begin{array}{r} 45 \\ \times 22 \\ \hline {}^1\,180 \\ 80 \\ \hline 990 \end{array} \qquad \begin{array}{r} 40+5 \\ \times 20+2 \\ \hline {}^{10} \\ 100 \quad 80+0 \\ 800+00 \\ \hline 900+90+0 \end{array}$$

18.
$$\begin{array}{r} 16 \\ \times 14 \\ \hline {}^2\,144 \\ 16 \\ \hline 224 \end{array} \qquad \begin{array}{r} 10+6 \\ \times 10+4 \\ \hline {}^{20} \\ 100+40+4 \\ 100 \quad 60+ \\ \hline 200+20+4 \end{array}$$

19.
$$\begin{array}{r} 39 \\ \times\;5 \\ \hline {}^{1\,4} \\ 55 \\ \hline 195 \end{array} \qquad \begin{array}{r} 30+9 \\ \times\quad 5 \\ \hline 100 \quad 40 \\ +50+5 \\ \hline 100+90+5 \end{array}$$

20. $30 \div 3 = 10$ yd
    $\$6 \times 10 = \$60$
21. $14 \times 18 = 252$ sq in
22. $46 + 28 = 74$ mi

Systematic Review 8F

1. $48 \div 6 = \underline{8}$
2. $18 \div 6 = \underline{3}$
3. $12 \div 6 = \underline{2}$
4. $36 \div 6 = \underline{6}$
5. $72 \div 9 = \underline{8}$
6. $54 \div 6 = \underline{9}$
7. $27 \div 3 = \underline{9}$
8. $45 \div 5 = \underline{9}$
9. $70 \div 10 = \underline{7}$
10. $16 \div 2 = \underline{8}$
11. $\dfrac{42}{6} = \underline{7}$
12. $\dfrac{60}{6} = \underline{10}$

13.
$$\begin{array}{r} {}^1 85 \\ +18 \\ \hline 103 \end{array}$$

14.
$$\begin{array}{r} {}^3\!\!\!\not{4}\,{}^1 7 \\ -\,3\,8 \\ \hline 9 \end{array}$$

15.
$$\begin{array}{r} {}^1 49 \\ +21 \\ \hline 70 \end{array}$$

16.
$$\begin{array}{r} {}^5\!\!\!\not{6}\,{}^1 4 \\ -\,2\,5 \\ \hline 3\,9 \end{array}$$

17.

$$
\begin{array}{r}
33 \\
\times 24 \\
\hline
{\scriptstyle 1} \\
122 \\
66 \\
\hline
792
\end{array}
\qquad
\begin{array}{r}
30+3 \\
\times 20+4 \\
\hline
{\scriptstyle 10} \\
100+20+2 \\
600+60 \\
\hline
700+90+2
\end{array}
$$

18.

$$
\begin{array}{r}
44 \\
\times 14 \\
\hline
{\scriptstyle 1\,1} \\
166 \\
44 \\
\hline
616
\end{array}
\qquad
\begin{array}{r}
40+4 \\
\times 10+4 \\
\hline
{\scriptstyle 100\quad 10} \\
100+60+6 \\
400+40 \\
\hline
600+10+6
\end{array}
$$

19.

$$
\begin{array}{r}
15 \\
\times 15 \\
\hline
{\scriptstyle 1\,2} \\
55 \\
15 \\
\hline
225
\end{array}
\qquad
\begin{array}{r}
10+5 \\
\times 10+5 \\
\hline
{\scriptstyle 100\quad 20} \\
+50+5 \\
100+50 \\
\hline
200+20+5
\end{array}
$$

20. $24 \div 6 = 4$ turns

21. $\$35 \times 14 = \$490$

22. $42 \div 6 = 7$ ft

Lesson Practice 9A

1. done

2. $4 \times 4 = 16$
$16 \div 2 = 8$ sq in

3. $2 \times 7 = 14$
$14 \div 2 = 7$ sq mi

4. $3 \times 6 = 18$
$18 \div 2 = 9$ sq ft

5. $4 \times 5 = 20$
$20 \div 2 = 10$ sq in

6. $9 \times 2 = 18$
$18 \div 2 = 9$ sq ft

7. $8 \times 2 = 16$
$16 \div 2 = 8$ sq yd

8. $1 \times 2 = 2$
$2 \div 2 = 1$ sq in

9. $2 \times 2 = 4$
$4 \div 2 = 2$ sq mi

10. $2 \times 4 = 8$
$8 \div 2 = 4$ sq in

Lesson Practice 9B

1. $3 \times 4 = 12$
$12 \div 2 = 6$ sq ft

2. $2 \times 6 = 12$
$12 \div 2 = 6$ sq in

3. $1 \times 8 = 8$
$8 \div 2 = 4$ sq mi

4. $2 \times 10 = 20$
$20 \div 2 = 10$ sq ft

5. $2 \times 5 = 10$
$10 \div 2 = 5$ sq in

6. $10 \times 1 = 10$
$10 \div 2 = 5$ sq in

7. $2 \times 3 = 6$
$6 \div 2 = 3$ sq yd

8. $3 \times 6 = 18$
$18 \div 2 = 9$ sq ft

9. $3 \div 3 = 1$ yd
$6 \div 3 = 2$ yd

10. $2 \times 1 = 2$
$2 \div 2 = 1$ sq yd

Lesson Practice 9C

1. $1 \times 4 = 4$
$4 \div 2 = 2$ sq ft

2. $2 \times 5 = 10$
$10 \div 2 = 5$ sq in

3. $2 \times 9 = 18$
$18 \div 2 = 9$ sq mi

4. $4 \times 4 = 16$
$16 \div 2 = 8$ sq ft

5. $2 \times 8 = 16$
$16 \div 2 = 8$ sq in

6. $2 \times 4 = 8$
$8 \div 2 = 4$ sq in

7. $2 \times 2 = 4$
$4 \div 2 = 2$ sq yd

8. $2 \times 7 = 14$
$14 \div 2 = 7$ sq ft

9. $4 \times 3 = 12$
$12 \div 2 = 6$ sq ft
6 plants

10. $6 \times 3 = 18$
$18 \div 2 = 9$ sq yd

Systematic Review 9D

1. $1 \times 4 = 4$
   $4 \div 2 = 2$ sq ft
2. $2 \times 3 = 6$
   $6 \div 2 = 3$ sq in
3. $5 \times 7 = 35$ sq in
4. $36 \div 6 = \underline{6}$
5. $42 \div 6 = \underline{7}$
6. $18 \div 6 = \underline{3}$
7. $54 \div 6 = \underline{9}$
8. $63 \div 9 = \underline{7}$
9. $40 \div 5 = \underline{8}$
10. $\frac{27}{3} = \underline{9}$
11. $\frac{80}{10} = \underline{8}$
12.
$$\begin{array}{r} 50 \\ \times 32 \\ \hline 100 \\ 150 \\ \hline 1600 \end{array} \qquad 50 \times 32 = \underline{1600}$$
13.
$$\begin{array}{r} 16 \\ \times 18 \\ \hline 4 \\ 88 \\ {}^{1}16 \\ \hline 288 \end{array} \qquad 16 \times 18 = \underline{288}$$
14.
$$\begin{array}{r} 28 \\ \times 22 \\ \hline {}^{1} \\ {}^{1}46 \\ 46 \\ \hline 616 \end{array} \qquad 28 \times 22 = \underline{616}$$
15.
$$\begin{array}{r} 32 \\ \times 17 \\ \hline {}^{1} \\ 214 \\ 32 \\ \hline 544 \end{array} \qquad 32 \times 17 = \underline{544}$$
16.
$$\begin{array}{r} {}^{1}23 \\ 26 \\ +37 \\ \hline 86 \end{array} \qquad 23 + 26 + 37 = 86$$
17.
$$\begin{array}{r} {}^{1}12 \\ 59 \\ +31 \\ \hline 102 \end{array} \qquad 12 + 59 + 31 = 102$$
18.
$$\begin{array}{r} {}^{1}15 \\ 15 \\ 44 \\ +24 \\ \hline 98 \end{array} \qquad 15 + 15 + 44 + 24 = 98$$
19.
$$\begin{array}{r} {}^{1}34 \\ 56 \\ 11 \\ +9 \\ \hline 110 \end{array} \qquad 34 + 56 + 11 + 9 = 110$$
20. $5 + 11 + 4 + 8 = 28$ pages

Systematic Review 9E

1. $6 \times 7 = 42$ sq ft
2. $2 \times 6 = 12$
   $12 \div 2 = 6$ sq in
3. $4 \times 8 = 32$ sq in
4. $30 \div 6 = \underline{5}$
5. $48 \div 6 = \underline{8}$
6. $12 \div 6 = \underline{2}$
7. $24 \div 6 = \underline{4}$
8. $72 \div 9 = \underline{8}$
9. $27 \div 9 = \underline{3}$
10. $\frac{35}{5} = \underline{7}$
11. $\frac{3}{3} = \underline{1}$
12.
$$\begin{array}{r} {}^{5}6\,{}^{1}0 \\ -3\,1 \\ \hline 2\,9 \end{array} \qquad 60 - 31 = 29$$
13.
$$\begin{array}{r} {}^{1}2\,{}^{1}7 \\ -1\,8 \\ \hline 9 \end{array} \qquad 27 - 18 = 9$$
14.
$$\begin{array}{r} {}^{4}5\,{}^{1}2 \\ -2\,7 \\ \hline 2\,5 \end{array} \qquad 52 - 27 = 25$$

15.
$$\begin{array}{r} {}^{1}16 \\ 72 \\ 38 \\ +31 \\ \hline 157 \end{array}$$
$16+72+38+31=157$

16.
$$\begin{array}{r} {}^{1}80 \\ 14 \\ 68 \\ 43 \\ +72 \\ \hline 277 \end{array}$$
$80+14+68+43+72=277$

17.
$$\begin{array}{r} 5 \\ {}^{3}39 \\ 84 \\ 71 \\ +26 \\ \hline 225 \end{array}$$
$5+39+84+71+26=225$

18. $54 \div 9 = 6$ teams

19. $7 \times 2 = 14$ pints

20. 2 sets

Systematic Review 9F

1. $5 \times 5 = 25$ sq ft

2. $9 \times 10 = 90$ sq ft

3. $3 \times 4 = 12$
$12 \div 2 = 6$ sq yd

4. $60 \div 6 = \underline{10}$

5. $36 \div 9 = \underline{4}$

6. $18 \div 3 = \underline{6}$

7. $\dfrac{20}{5} = \underline{4}$

8. $45 \times 16 = 720$

9. $52 \times 28 = 1,456$

10. $76 \times 54 = 4,104$

11. $33+75+44+67 = 219$

12. $83+90+45+25+17 = 260$

13. $26+43+31+57+14 = 171$

14. $\$45 \div \$5 = 9$
$9 - 1$ (himself) $= 8$ friends

15. $21 \div 3 = 7$ boxes

16. $42 \div 6 = 7$ treats

17. $11+9+13 = 33$ points

18. $62 - 49 = 13\text{¢}$

19. $12 \times 25 = 300$ cookies

20. no

Lesson Practice 10A

1. 4,8,12,16,20;$\underline{5}$

2. 4,8;$\underline{2}$

3. 4,8,12,16,20,24,28;$\underline{7}$

4. done

5.
$$\begin{array}{r} 10 \\ 4\overline{)40} \\ \underline{-40} \\ 0 \end{array}$$

6.
$$\begin{array}{r} 1 \\ 4\overline{)4} \\ \underline{-4} \\ 0 \end{array}$$

7.
$$\begin{array}{r} 5 \\ 4\overline{)20} \\ \underline{-20} \\ 0 \end{array}$$

8.
$$\begin{array}{r} 8 \\ 4\overline{)32} \\ \underline{-32} \\ 0 \end{array}$$

9.
$$\begin{array}{r} 7 \\ 4\overline{)28} \\ \underline{-28} \\ 0 \end{array}$$

10. $12 \div 4 = \underline{3}$

11. $36 \div 4 = \underline{9}$

12. $40 \div 4 = \underline{10}$

13. $\dfrac{16}{4} = \underline{4}$

14. $\dfrac{32}{4} = \underline{8}$

15. $36 \div 4 = 9$

16. $24 \div 4 = 6$ cups

17. $28 \div 4 = 7$ days

18. $16 \div 4 = 4$ chairs

Lesson Practice 10B

1. 4,8,12,16,20,24,28,32,36;<u>9</u>
2. 4;<u>1</u>
3. 4,8,12;<u>3</u>
4. $4\overline{)32}$ → 8, −32, 0
5. $4\overline{)16}$ → 4, −16, 0
6. $4\overline{)8}$ → 2, −8, 0
7. $4\overline{)24}$ → 6, −24, 0
8. $4\overline{)12}$ → 3, −12, 0
9. $4\overline{)36}$ → 9, −36, 0
10. $20 \div 4 = \underline{5}$
11. $28 \div 4 = \underline{7}$
12. $32 \div 4 = \underline{8}$
13. $\frac{40}{4} = \underline{10}$
14. $\frac{4}{4} = \underline{1}$
15. $24 \div \underline{4} = 6$
16. $36 \div 4 = 9$ cards
17. $\$20 \div \$4 = 5$ hours
18. $40 \div 4 = 10$ horses

Lesson Practice 10C

1. 4,8,12,16,20,24;<u>6</u>
2. 4,8,12,16;<u>4</u>
3. 4,8,12,16,20,24,28,32;<u>8</u>
4. $4\overline{)8}$ → 2, −8, 0
5. $4\overline{)28}$ → 7, −28, 0
6. $4\overline{)40}$ → 10, −40, 0
7. $4\overline{)36}$ → 9, −36, 0
8. $4\overline{)20}$ → 5, −20, 0
9. $4\overline{)4}$ → 1, −4, 0
10. $12 \div 4 = \underline{3}$
11. $24 \div 4 = \underline{6}$
12. $16 \div 4 = \underline{4}$
13. $\frac{8}{4} = \underline{2}$
14. $\frac{28}{4} = \underline{7}$
15. $\frac{40}{4} = \underline{10}$
16. $32 \div 4 = 8$ cages
17. $28 \div 4 = 7$ chocolates
18. $8 \div 4 = 2$ years

Systematic Review 10D

1.
$$\begin{array}{r} 9 \\ 4\overline{)36} \\ -36 \\ \hline 0 \end{array}$$

2.
$$\begin{array}{r} 5 \\ 4\overline{)20} \\ -20 \\ \hline 0 \end{array}$$

3.
$$\begin{array}{r} 4 \\ 4\overline{)16} \\ -16 \\ \hline 0 \end{array}$$

4.
$$\begin{array}{r} 7 \\ 4\overline{)28} \\ -28 \\ \hline 0 \end{array}$$

5. $36 \div 6 = \underline{6}$
6. $14 \div 2 = \underline{7}$
7. $\frac{15}{3} = \underline{5}$
8. $\frac{81}{9} = \underline{9}$

9.
$$\begin{array}{r} {}^{1}13 \\ 25 \\ 37 \\ +42 \\ \hline 117 \end{array}$$

10.
$$\begin{array}{r} {}^{2}\cancel{3}\,{}^{1}1 \\ -22 \\ \hline 9 \end{array}$$

11.
$$\begin{array}{r} {}^{4}\cancel{5}\,{}^{1}8 \\ -39 \\ \hline 19 \end{array}$$

12. $45 \times 15 = 675$
13. $25 \times 12 = 300$ sq ft
14. $12 \times 6 = 72$ sq in
15. $2 \times 5 = 10$
    $10 \div 2 = 5$ sq in
16. $8 \times 4 = 32$ quarters
17. $8 \div 4 = 2$ gallons
18. $6 \times 4 = 24$ quarts

Systematic Review 10E

1.
$$\begin{array}{r} 10 \\ 4\overline{)40} \\ -40 \\ \hline 0 \end{array}$$

2.
$$\begin{array}{r} 3 \\ 4\overline{)12} \\ -12 \\ \hline 0 \end{array}$$

3.
$$\begin{array}{r} 8 \\ 4\overline{)32} \\ -32 \\ \hline 0 \end{array}$$

4.
$$\begin{array}{r} 6 \\ 4\overline{)24} \\ -24 \\ \hline 0 \end{array}$$

5. $100 \div 10 = \underline{10}$
6. $35 \div 5 = \underline{7}$
7. $\frac{27}{9} = \underline{3}$
8. $\frac{54}{6} = \underline{9}$

9.
$$\begin{array}{r} {}^{2} \\ 46 \\ 14 \\ 23 \\ +17 \\ \hline 100 \end{array}$$

10.
$$\begin{array}{r} {}^{6}\cancel{7}\,{}^{1}6 \\ -47 \\ \hline 29 \end{array}$$

11. $64 \times 32 = 2{,}048$
12. $43 \times 84 = 3{,}612$
13. $5 \times 4 = 20$
    $20 \div 2 = 10$ sq ft
14. $23 \times 28 = 644$ sq mi
15. $3 \times 6 = 18$
    $18 \div 2 = 9$ sq ft
16. $\$28 \div 4 = \$7$
17. $20 \div 4 = \$5$
18. $24 \times 12 = 288$ books
19. $45 \div 5 = 9$ shots

20. $6 + 8 + 10 + 12 = 36$ quarts
$36 \div 4 = 9$ gallons

Systematic Review 10F

1.
$$\begin{array}{r} 4 \\ 4\overline{)16} \\ \underline{-16} \\ 0 \end{array}$$

2.
$$\begin{array}{r} 1 \\ 4\overline{)4} \\ \underline{-4} \\ 0 \end{array}$$

3.
$$\begin{array}{r} 10 \\ 4\overline{)40} \\ \underline{-40} \\ 0 \end{array}$$

4.
$$\begin{array}{r} 2 \\ 4\overline{)8} \\ \underline{-8} \\ 0 \end{array}$$

5. $18 \div 2 = \underline{9}$

6. $24 \div 3 = \underline{8}$

7. $\dfrac{42}{6} = \underline{7}$

8. $\dfrac{72}{9} = \underline{8}$

9.
$$\begin{array}{r} {}^2 3\ 8 \\ 4\ 1 \\ 1\ 2 \\ +\ \ \ 9 \\ \hline 1\ 0\ 0 \end{array}$$

10.
$$\begin{array}{r} {}^1 5\ 6 \\ 2\ 4 \\ 1\ 8 \\ +\ 2\ 1 \\ \hline 1\ 1\ 9 \end{array}$$

11.
$$\begin{array}{r} {}^8 9\ {}^1\!\! 1 \\ -\ 2\ 7 \\ \hline 6\ 4 \end{array}$$

12.
$$\begin{array}{r} 7\ 5 \\ -2\ 5 \\ \hline 5\ 0 \end{array}$$

13. $7 \times 2 = 14$
$14 \div 2 = 7$ sq ft

14. $16 \times 9 = 144$ sq mi

15. $2 \times 4 = 8$
$8 \div 2 = 4$ sq in

16. $12 \div 4 = 3$ packs

17. $32 \div 4 = \$8$

18. $46 \times 46 = 2,116$ sq ft

19. $16 + 19 + 10 + 15 = 60$ fireflies
$60 \times 2 = 120$ minutes

20. $24 \div 4 = 6$ teams

Lesson Practice 11A

1. done

2. $5 + 4 + 6 = 15$
$15 \div 3 = \underline{5}$

3. $8 + 8 + 5 = 21$
$21 \div 3 = \underline{7}$

4. $9 + 11 = 20$
$20 \div 2 = \underline{10}$

5. $2 + 3 + 1 = 6$
$6 \div 3 = \underline{2}$

6. $10 + 6 = 16$
$16 \div 2 = \underline{8}$

7. $10 + 7 + 6 + 9 = 32$
$32 \div 4 = \underline{8}$

8. $2 + 8 + 5 + 9 + 1 = 25$
$25 \div 5 = \underline{5}$

9. $7 + 6 + 14 + 9 = 36$
$36 \div 4 = \underline{9}$

10. $9 + 15 + 6 = 30$
$30 \div 3 = 10$ cards per month

11. $4 + 5 + 3 + 4 + 4 = 20$
$20 \div 5 = 4$ emails per day

12. $8 + 6 + 5 + 10 + 13 + 12 = 54$
$54 \div 6 = 9$ pages per day

Lesson Practice 11B

1. $2 + 5 + 5 = 12$
$12 \div 3 = \underline{4}$

2. $10 + 5 + 9 = 24$
$24 \div 3 = \underline{8}$

3. $11 + 6 + 10 = 27$
$27 \div 3 = \underline{9}$

4. $2 + 8 = 10$
$10 \div 2 = \underline{5}$

5. $7 + 9 + 14 + 6 = 36$
$36 \div 4 = \underline{9}$

6. $7 + 5 = 12$
$12 \div 2 = \underline{6}$

7. $10 + 3 + 6 + 1 = 20$
$20 \div 4 = \underline{5}$

8. $8 + 5 + 8 + 7 = 28$
$28 \div 4 = \underline{7}$

9. $3 + 3 + 6 + 9 + 9 = 30$
$30 \div 5 = \underline{6}$

10. $10 + 11 + 9 + 10 = 40$
$40 \div 4 = 10$ jewels per mine

11. $8 + 7 + 10 \div 7 = 32$
$32 \div 4 = 8$ parts per vehicle

12. $8 + 7 + 4 + 10 + 9 + 10 = 48$
$48 \div 6 = 8$ in per month

Lesson Practice 11C

1. $8 + 10 + 12 = 30$
$30 \div 3 = \underline{10}$

2. $7 + 6 + 8 = 21$
$21 \div 3 = \underline{7}$

3. $1 + 2 + 6 = 9$
$9 \div 3 = \underline{3}$

4. $1 + 5 = 6$
$6 \div 2 = \underline{3}$

5. $1 + 3 + 4 + 8 = 16$
$16 \div 4 = \underline{4}$

6. $6 + 8 = 14$
$14 \div 2 = \underline{7}$

7. $1 + 3 + 6 + 7 + 13 = 30$
$30 \div 5 = \underline{6}$

8. $4 + 2 + 6 + 8 = 20$
$20 \div 4 = \underline{5}$

9. $9 + 12 + 2 + 5 = 28$
$28 \div 4 = \underline{7}$

10. $9 + 12 + 8 + 11 = 40$
$40 \div 4 = 10$ points per quarter

11. $10 + 9 + 12 + 5 + 9 = 45$
$45 \div 5 = 9$ tickets

12. $2 + 3 + 4 + 6 + 8 + 1 = 24$
$24 \div 6 = 4$ in

Systematic Review 11D

1. $2 + 1 + 3 + 6 = 12$
$12 \div 4 = \underline{3}$

2. $5 + 7 + 3 + 10 + 15 = 40$
$40 \div 5 = \underline{8}$

3. $9 + 4 + 5 = 18$
$18 \div 3 = \underline{6}$

4. $7 \times \underline{7} = 49$

5. $8 \times \underline{6} = 48$

6. $7 \times \underline{8} = 56$

7. $\begin{array}{r} 4 \\ 9\overline{)36} \\ -36 \\ \hline 0 \end{array}$

8. $\begin{array}{r} 10 \\ 5\overline{)50} \\ -50 \\ \hline 0 \end{array}$

9. $\begin{array}{r} 7 \\ 6\overline{)42} \\ -42 \\ \hline 0 \end{array}$

10. $18 \div 2 = \underline{9}$

11. $63 \div 9 = \underline{7}$

12. $\dfrac{30}{6} = \underline{5}$

13. $\begin{array}{r} {}^6\cancel{7}\,{}^1 2 \\ -\;3\;4 \\ \hline 3\;8 \end{array}$

14. $15 \times 47 = 705$

15. $28 \times 18 = 504$

16. $7 \times 8 = 56$ sq ft

17. $2 \times 8 = 16$
$16 \div 2 = 8$ sq in

18. $3+2+3+5+2=15$
$15\div5=3$ hours per day
19. $24\div4=6$ gallons
20. perpendicular

18. $36\div4=\$9$
19. $29+25=54$
$54\div9=6$ stickers each
20. $8\times\underline{7}=56$

Systematic Review 11E
1. $9+6+7+2=24$
$24\div4=\underline{6}$
2. $1+3+9+8+7+8=36$
$36\div6=\underline{6}$
3. $8+6+13=27$
$27\div3=\underline{9}$
4. $8\times\underline{8}=64$
5. $7\times\underline{9}=63$
6. $8\times\underline{5}=40$
7.
$$9\overline{)27}$$ quotient 3, -27, 0
8.
$$9\overline{)72}$$ quotient 8, -72, 0
9.
$$2\overline{)4}$$ quotient 2, -4, 0
10. $45\div9=\underline{5}$
11. $18\div6=\underline{3}$
12. $\frac{70}{10}=\underline{7}$
13. $37\times17=629$
14. $48-21=27$
15. $53-29=24$
16. $10\times12=120$ sq ft
17. $4\times3=12$
$12\div2=6$ sq ft

Systematic Review 11F
1. $3+4+5+4=16$
$16\div4=\underline{4}$
2. $1+1+2+3+3=10$
$10\div5=\underline{2}$
3. $4+7+4=15$
$15\div3=\underline{5}$
4. $8\times\underline{9}=72$
5. $7\times\underline{8}=56$
6. $7\times\underline{7}=49$
7.
$$9\overline{)18}$$ quotient 2, -18, 0
8.
$$6\overline{)12}$$ quotient 2, -12, 0
9.
$$6\overline{)54}$$ quotient 9, -54, 0
10. $25\div5=\underline{5}$
11. $80\div10=\underline{8}$
12. $\frac{81}{9}=\underline{9}$
13. $55\times27=1,485$
14. $62\times38=2,356$
15. $95-46=49$
16. $8\times8=64$ sq in
17. $1\times6=6$
$6\div2=3$ sq mi
18. $30\div3=10$ yd

19. $9 \times 3 = 27$
$27 - 5 = 22$ ft
20. E,F,H,M,N,Z
(possibly I, depending on style)

17. $14 \div 7 = 2$ cookies
18. $48 \div 8 = 6$ octopuses

Lesson Practice 12A

1. 8,16,24;3
2. 7,14,21,28,35,42,49;7
3. 8,16;2
4.
$$\begin{array}{r} 8 \\ 7\overline{)56} \\ -56 \\ \hline 0 \end{array}$$
5.
$$\begin{array}{r} 8 \\ 8\overline{)64} \\ -64 \\ \hline 0 \end{array}$$
6.
$$\begin{array}{r} 6 \\ 7\overline{)42} \\ -42 \\ \hline 0 \end{array}$$
7.
$$\begin{array}{r} 5 \\ 8\overline{)40} \\ -40 \\ \hline 0 \end{array}$$
8.
$$\begin{array}{r} 4 \\ 8\overline{)32} \\ -32 \\ \hline 0 \end{array}$$
9.
$$\begin{array}{r} 9 \\ 7\overline{)63} \\ -63 \\ \hline 0 \end{array}$$
10. $56 \div 8 = 7$
11. $35 \div 7 = 5$
12. $48 \div 8 = 6$
13. $\frac{28}{7} = 4$
14. $\frac{21}{7} = 3$
15. $72 \div 8 = 9$
16. $56 \div 8 = 7$ weeks

Lesson Practice 12B

1. 7,14,21,28,35,42,49,56,63;9
2. 8,16,24,32,40,48,56,64;8
3. 7,14,21,28,35,42;6
4.
$$\begin{array}{r} 10 \\ 7\overline{)70} \\ -70 \\ \hline 0 \end{array}$$
5.
$$\begin{array}{r} 3 \\ 8\overline{)24} \\ -24 \\ \hline 0 \end{array}$$
6.
$$\begin{array}{r} 7 \\ 7\overline{)49} \\ -49 \\ \hline 0 \end{array}$$
7.
$$\begin{array}{r} 6 \\ 8\overline{)48} \\ -48 \\ \hline 0 \end{array}$$
8.
$$\begin{array}{r} 1 \\ 8\overline{)8} \\ -8 \\ \hline 0 \end{array}$$
9.
$$\begin{array}{r} 8 \\ 7\overline{)56} \\ -56 \\ \hline 0 \end{array}$$
10. $40 \div 8 = 5$
11. $28 \div 7 = 4$
12. $72 \div 8 = 9$
13. $\frac{21}{7} = 3$
14. $\frac{35}{7} = 5$
15. $56 \div 8 = 7$
16. $32 \div 8 = 4$ hours

17. $56 \div 7 = 8$ weeks
18. $63 \div 7 = 9$ in

Lesson Practice 12C

1. 8,16,24,32,40,48,56;<u>7</u>
2. 7,14,21,28,35,42,49,56;<u>8</u>
3. 8,16,24,32,40,48,56,64,72;<u>9</u>
4.
$$\begin{array}{r} 9 \\ 7\overline{)63} \\ \underline{-63} \\ 0 \end{array}$$
5.
$$\begin{array}{r} 4 \\ 8\overline{)32} \\ \underline{-32} \\ 0 \end{array}$$
6.
$$\begin{array}{r} 3 \\ 7\overline{)21} \\ \underline{-21} \\ 0 \end{array}$$
7.
$$\begin{array}{r} 8 \\ 8\overline{)64} \\ \underline{-64} \\ 0 \end{array}$$
8.
$$\begin{array}{r} 3 \\ 8\overline{)24} \\ \underline{-24} \\ 0 \end{array}$$
9.
$$\begin{array}{r} 7 \\ 7\overline{)49} \\ \underline{-49} \\ 0 \end{array}$$
10. $16 \div 8 = \underline{2}$
11. $42 \div 7 = \underline{6}$
12. $40 \div 8 = \underline{5}$
13. $\frac{35}{7} = \underline{5}$
14. $\frac{14}{7} = \underline{2}$
15. $\frac{80}{8} = \underline{10}$
16. $70 \div 7 = 10$ troops
17. $48 \div 8 = 6$ words
18. $28 \div 7 = 4$ miles

Systematic Review 12D

1.
$$\begin{array}{r} 8 \\ 7\overline{)56} \\ \underline{-56} \\ 0 \end{array}$$
2.
$$\begin{array}{r} 8 \\ 8\overline{)64} \\ \underline{-64} \\ 0 \end{array}$$
3.
$$\begin{array}{r} 7 \\ 8\overline{)56} \\ \underline{-56} \\ 0 \end{array}$$
4.
$$\begin{array}{r} 7 \\ 7\overline{)49} \\ \underline{-49} \\ 0 \end{array}$$
5. $16 \div 4 = \underline{4}$
6. $36 \div 6 = \underline{6}$
7. $\frac{72}{9} = \underline{8}$
8. $\frac{21}{7} = \underline{3}$
9. $7 + 1 + 2 + 6 + 4 = 20$
   $20 \div 5 = \underline{4}$
10. $10 + 13 + 9 + 9 + 8 + 11 = 60$
    $60 \div 6 = 10$
11. $6 + 5 + 13 = 24$
    $24 \div 3 = 8$
12.
$$\begin{array}{r} 1\overset{1}{2}4 \\ +369 \\ \hline 493 \end{array}$$
13.
$$\begin{array}{r} \overset{1}{7}\overset{1}{8}1 \\ +319 \\ \hline 1100 \end{array}$$
14.
$$\begin{array}{r} 3\overset{1}{3}5 \\ +126 \\ \hline 461 \end{array}$$

15.
$$\begin{array}{r} 4\,{}^{1}0\,4 \\ +2\,7\,8 \\ \hline 6\,8\,2 \end{array}$$

16. $8 \times 16 = 128$ oz

17. $36 \div 4 = \$9$

18. $3 + 4 + 2 + 3 = 12$
$12 \div 4 = 3$ hours

19. $9 \times 9 = 81$ sq in

20. $18 \times 3 = 54$ ft

Systematic Review 12E

1.
$$\begin{array}{r} 4 \\ 7\,\overline{)28} \\ -28 \\ \hline 0 \end{array}$$

2.
$$\begin{array}{r} 6 \\ 8\,\overline{)48} \\ -48 \\ \hline 0 \end{array}$$

3.
$$\begin{array}{r} 9 \\ 7\,\overline{)63} \\ -63 \\ \hline 0 \end{array}$$

4.
$$\begin{array}{r} 6 \\ 7\,\overline{)42} \\ -42 \\ \hline 0 \end{array}$$

5. $35 \div 5 = \underline{7}$

6. $27 \div 3 = \underline{9}$

7. $\dfrac{54}{9} = \underline{6}$

8. $\dfrac{56}{7} = \underline{8}$

9. $9 + 4 + 5 + 6 = 24$
$24 \div 4 = \underline{6}$

10. $1 + 2 + 3 + 4 + 4 + 5 + 6 + 7 = 32$
$32 \div 8 = \underline{4}$

11. $4 + 5 + 9 = 18$
$18 \div 3 = \underline{6}$

12.
$$\begin{array}{r} 4\,{}^{1}2\,{}^{1}1 \\ -2\,0\,8 \\ \hline 2\,1\,3 \end{array}$$

13.
$$\begin{array}{r} 6\,{}^{3}4\,{}^{1}2 \\ -1\,2\,7 \\ \hline 5\,1\,5 \end{array}$$

14.
$$\begin{array}{r} {}^{6}7\,{}^{1}8\,9 \\ -\,3\,9\,4 \\ \hline 3\,9\,5 \end{array}$$

15.
$$\begin{array}{r} {}^{2}3\,{}^{1}0\,3 \\ -\,1\,6\,3 \\ \hline 1\,4\,0 \end{array}$$

16. $2 \times 9 = 18$
$18 \div 2 = 9$ sq ft

17. $25 \times 16 = 400$ oz

18. $1 + 5 + 10 + 4 = 20$
$20 \div 4 = 5$ mistakes

19. $251 + 317 = 568$
$568 - 179 = 389$ nuts

20. $49 \div 7 = 7$ books

Systematic Review 12F

1.
$$\begin{array}{r} 5 \\ 7\,\overline{)35} \\ -35 \\ \hline 0 \end{array}$$

2.
$$\begin{array}{r} 9 \\ 8\,\overline{)72} \\ -72 \\ \hline 0 \end{array}$$

3.
$$\begin{array}{r} 2 \\ 7\,\overline{)14} \\ -14 \\ \hline 0 \end{array}$$

4.
$$\begin{array}{r} 3 \\ 7\,\overline{)21} \\ -21 \\ \hline 0 \end{array}$$

5. $40 \div 8 = \underline{5}$

6. $16 \div 8 = \underline{2}$

7. $\dfrac{45}{5} = \underline{9}$

8. $\dfrac{64}{8} = \underline{8}$

9. $2+4+6+8+10=30$
$30 \div 5 = 6$
10. $1+1+2+2+4+4+5+5=24$
$24 \div 8 = \underline{3}$
11. $4+10+16=30$
$30 \div 3 = \underline{10}$
12. $17 \times 25 = \underline{425}$
13. $48 \times 36 = \underline{1,728}$
14. $89 \times 43 = \underline{3,827}$
15. $78 \times 87 = \underline{6,786}$
16. $5 \times 9 = 45$ sq in
17. 1 lb = 16 oz
20 oz > 16 oz
18. $18 \div 2 = 9$ jars
19. 32 qt ÷ 4 = 8 gal
8 gal – 2 gal = 6 gal
20. $12 \times 24 = 288$ pencils

Lesson Practice 13A
1. done
2. $9+11=20$
$20 \div 2 = 10$
$10 \times 6 = 60$ sq in
3. $6+10=16$
$16 \div 2 = 8$
$8 \times 4 = 32$ sq ft
4. $6+12=18$
$18 \div 2 = 9$
$9 \times 7 = 63$ sq ft
5. $3+5=8$
$8 \div 2 = 4$
$4 \times 2 = 8$ sq ft
6. $8+10=18$
$18 \div 2 = 9$
$9 \times 5 = 45$ sq in
7. $7+9=16$
$16 \div 2 = 8$
$8 \times 5 = 40$ sq in
8. $1+3=4$
$4 \div 2 = 2$
$2 \times 2 = 4$ sq mi

Lesson Practice 13B
1. $2+4=6$
$6 \div 2 = 3$
$3 \times 3 = 9$ sq in
2. $3+17=20$
$20 \div 2 = 10$
$10 \times 6 = 60$ sq ft
3. $5+9=14$
$14 \div 2 = 7$
$7 \times 5 = 35$ sq ft
4. $4+10=14$
$14 \div 2 = 7$
$7 \times 3 = 21$ sq in
5. $8+12=20$
$20 \div 2 = 10$
$10 \times 11 = 110$ sq ft
6. $7+11=18$
$18 \div 2 = 9$
$9 \times 4 = 36$ sq in
7. $2+6=8$
$8 \div 2 = 4$
$4 \times 6 = 24$ sq ft
24 plants
8. $5+7=12$
$12 \div 2 = 6$
$6 \times 6 = 36$ sq in

Lesson Practice 13C
1. $4+10=14$
$14 \div 2 = 7$
$7 \times 4 = 28$ sq in
2. $1+11=12$
$12 \div 2 = 6$
$6 \times 3 = 18$ sq ft
3. $6+8=14$
$14 \div 2 = 7$
$7 \times 5 = 35$ sq ft
4. $4+6=10$
$10 \div 2 = 5$
$5 \times 4 = 20$ sq in
5. $5+11=16$
$16 \div 2 = 8$
$8 \times 10 = 80$ sq ft

6. $2+14=16$
$16\div2=8$
$8\times6=48$ sq ft

7. $1+5=6$
$6\div2=3$
$3\times2=6$ sq mi

8. $6+14=20$
$20\div2=10$
$10\times15=150$ sq ft

Systematic Review 13D

1. $5+7=12$
$12\div2=6$
$6\times6=36$ sq in

2. $3+15=18$
$18\div2=9$
$9\times5=45$ sq ft

3. $27\div9=\underline{3}$

4. $36\div6=\underline{6}$

5. $28\div7=\underline{4}$

6. $45\div5=\underline{9}$

7. $56\div8=\underline{7}$

8. $49\div7=\underline{7}$

9. $\dfrac{16}{2}=\underline{8}$

10. $\dfrac{42}{6}=\underline{7}$

11. $6+5+7+10=28$
$28\div4=\underline{7}$

12. $2+6+1+4+3+2=18$
$18\div6=\underline{3}$

13. $1+5+15=21$
$21\div3=\underline{7}$

14. $7\times20=\underline{140}$

15. $13\times20=\underline{260}$

16. $9\times30=\underline{270}$

17. $24\div4=6$ horses

18. $60\div6=10$ songs

19. $20\times12=240$ months

20. $\$90-\$65=\$25$

Systematic Review 13E

1. $3+11=14$
$14\div2=7$
$7\times7=49$ sq ft

2. $25\times18=450$ sq in

3. $63\div9=\underline{7}$

4. $72\div9=\underline{8}$

5. $2\div2=\underline{1}$

6. $18\div3=\underline{6}$

7. $32\div4=\underline{8}$

8. $25\div5=\underline{5}$

9. $\dfrac{48}{6}=\underline{8}$

10. $\dfrac{56}{7}=\underline{8}$

11.
$$\begin{array}{r} {}^{1}1^{1}78 \\ +3\ 34 \\ \hline 5\ 12 \end{array}$$

12.
$$\begin{array}{r} 5^{1}\!\not{2}\,^{1}\!1 \\ -4\ \ 13 \\ \hline 1\ 08 \end{array}$$

13.
$$\begin{array}{r} 7^{1}08 \\ +2\ 22 \\ \hline 9\ 30 \end{array}$$

14. $9\times20=\underline{180}$

15. $34\times20=\underline{680}$

16. $11\times30=\underline{330}$

17. $63\div7=9$ days

18. $24\div4=6$ ft

19. $11+15+4+10+5=45$ dogs
$45\div5=9$ dogs

20. $13\times16=208$ oz

Systematic Review 13F

1. $2+18=20$
$20\div2=10$
$10\times6=60$ sq ft

2. $2\times3=6$
$6\div2=3$ sq ft

3. $35\div7=\underline{5}$

4. $36\div9=\underline{4}$

5. $54\div6=\underline{9}$

6. $64 \div 8 = \underline{8}$
7. $48 \div 8 = \underline{6}$
8. $24 \div 6 = \underline{4}$
9. $\dfrac{50}{5} = \underline{10}$
10. $\dfrac{21}{7} = \underline{3}$
11. $16 \div 4 = \underline{4}$
12. $36 \div 4 = \underline{9}$
13. $24 \div 3 = \underline{8}$
14. $6 \times 30 = \underline{180}$
15. $21 \times 20 = \underline{420}$
16. $42 \times 30 = \underline{1,260}$
17. $32 \div 8 = 4$ cakes
18. $20 \div 4 = \$5$
19. $50 - 20 = 30$ quarters
    $6 \times 4 = 24$ quarters
    He needs enough for 30
    but has only enough for 24.
20. $419 + 495 = 914$ words

Lesson Practice 14A

1. done
2. 35,201
3. 765,892
4. 4,265,143
5. done
6. $10,000 + 100 + 20 + 9$
7. $100,000 + 90,000 + 5,000 +$
    $300 + 20 + 8$
8. $1,000,000 + 700,000 + 80,000 +$
    $6,000 + 200 + 1$
9. done
10. 28,616
11. 4,300,400
12. 6,815,231

Lesson Practice 14B

1. 2,794
2. 16,302
3. 651,741
4. 2,540,000

5. $7,000 + 800 + 1$
6. $40,000 + 1,000 + 400 + 50 + 6$
7. $200,000 + 30,000 + 8,000 + 100 + 90 + 9$
8. $5,000,000 + 300,000 + 60,000 + 5,000$
9. 3,021
10. 45,615
11. 5,400,000
12. 8,131,528

Lesson Practice 14C

1. 1,224
2. 43,638
3. 247,000
4. 3,122,472
5. $3,000 + 200 + 50 + 6$
6. $50,000 + 600 + 4$
7. $700,000 + 50,000 + 4,000 +$
    $700 + 50 + 3$
8. $2,000,000 + 100,000 + 10,000 +$
    $7,000 + 200 + 40 + 9$
9. 1,838
10. 33,230
11. 2,350,000
12. 4,652,893

Systematic Review 14D

1. 1,652
2. $6,000,000 + 300,000 + 40,000 +$
    $100 + 20 + 9$
3. $4 + 12 = 16$
    $16 \div 2 = 8$
    $8 \times 9 = 72$ sq ft
4. $2 \times 6 = 12$
    $12 \div 2 = 6$ sq in
5. $42 \div 7 = \underline{6}$
6. $54 \div 6 = \underline{9}$
7. $81 \div 9 = \underline{9}$
8. $40 \div 8 = \underline{5}$
9. $14 \div 7 = \underline{2}$
10. $12 \div 4 = \underline{3}$
11. $\dfrac{15}{3} = \underline{5}$

12. $\frac{70}{7} = \underline{10}$

13. done

14. $17 \times 100 = 1,700$

15. $8 \times 200 = 1,600$

16. $478 - 182 = 296$ marbles

17. $555 + 555 = 1,110$ ft

18. $24 \div 4 = 6$ cookies

Systematic Review 14E

1. 25,611

2. $1,000,000 + 100,000 + 70,000 + 4,000$

3. $2 + 4 = 6$
   $6 \div 2 = 3$
   $3 \times 3 = 9$ sq in

4. $12 \times 12 = 144$ sq mi

5. $35 \div 5 = \underline{7}$

6. $18 \div 6 = \underline{3}$

7. $28 \div 7 = \underline{4}$

8. $27 \div 9 = \underline{3}$

9. $72 \div 8 = \underline{9}$

10. $20 \div 5 = \underline{4}$

11. $\frac{8}{4} = \underline{2}$

12. $\frac{12}{3} = \underline{4}$

13. $23 \times 100 = \underline{2,300}$

14. $14 \times 200 = \underline{2,800}$

15. $9 \times 200 = \underline{1,800}$

16. $8 \times 11 = 88$ sq ft

17. $12 \times 10 = 120$ logs

18. $14 - 6 = 8$ qt
    $8 \times 2 = 16$ pt

Systematic Review 14F

1. 14,715

2. $4,000,000 + 700,000 + 10,000 + 1,000 + 300 + 40$

3. $9 + 11 = 20$
   $20 \div 2 = 10$
   $10 \times 10 = 100$ sq in

4. $23 \times 38 = 874$ sq ft

5. $90 \div 10 = \underline{9}$

6. $18 \div 9 = \underline{2}$

7. $49 \div 7 = \underline{7}$

8. $42 \div 6 = \underline{7}$

9. $14 \div 2 = \underline{7}$

10. $7 \div 1 = \underline{7}$

11. $\frac{36}{6} = \underline{6}$

12. $\frac{28}{4} = \underline{7}$

13. $7 \times 200 = \underline{1,400}$

14. $33 \times 200 = \underline{6,600}$

15. $15 \times 100 = \underline{1,500}$

16. parallel

17. $15 \times 16 = 240$ oz

18. $7 + 15 + 13 + 9 + 6 = 50$
    $50 \div 5 = 10$ points

Lesson Practice 15A

1. done

2. 11,368; eleven thousand, three hundred sixty-eight

3. 3,000,000; three million

4. 6,000,000,000; six billion

5. done

6. $3 \times 10,000 + 1 \times 100 + 8 \times 1$

7. $2 \times 1,000,000,000 + 5 \times 100,000,000 + 4 \times 10,000,000 + 2 \times 1,000,000$

8. $6 \times 1,000,000,000,000,000$

9. done

10. 2,113,000

11. 7,945,000,000

12. 2,000,000,000,000,112

Lesson Practice 15B

1. 50,940; fifty thousand, nine hundred forty

2. 672,800; six hundred seventy-two thousand, eight hundred

3. 94,000,000; ninety-four million

4. 2,648,000,000,000; two trillion, six hundred forty-eight billion

5. $5 \times 100,000,000,000,000 + 1 \times 10 + 9 \times 1$

6. $1 \times 1,000,000,000,000 + 7 \times 100,000,000,000 + 8 \times 10,000,000,000 + 3 \times 1,000,000,000$

7. $7 \times 10,000,000,000 + 2 \times 1,000,000,000 + 3 \times 100,000,000 + 5 \times 10,000,000$

8. $7 \times 1,000,000,000,000 + 6 \times 100 + 2 \times 10$

9. 59,140

10. 1,000,089

11. 32,000,477,000

12. 7,891,000,000,000

Lesson Practice 15C

1. 60,000,006; sixty million, six

2. 495,006,200; four hundred ninety-five million, six thousand, two hundred

3. 36,000,000; thirty-six million

4. 3,003,000,000,000; three trillion, three billion

5. $7 \times 100,000,000,000 + 2 \times 10,000,000,000 + 5 \times 1,000,000,000 + 7 \times 10,000,000 + 8 \times 1,000,000$

6. $4 \times 10,000,000,000,000 + 3 \times 100,000,000,000 + 1 \times 10,000,000,000 + 6 \times 1,000,000,000$

7. $1 \times 1,000,000,000 + 4 \times 100,000,000 + 6 \times 10,000,000 + 5 \times 1,000,000 + 9 \times 100,000$

8. $3 \times 100,000,000,000,000 + 7 \times 10,000,000,000,000 + 1 \times 1,000,000,000,000 + 8 \times 100,000,000,000$

9. 256,094

10. 6,000,851,000,000

11. 874,000,320

12. 1,067,000,000

Systematic Review 15D

1. 547,000,000,000; five hundred forty-seven billion

2. $5 \times 100,000,000,000,000 + 6 \times 10,000,000,000,000 + 4 \times 1,000,000,000,000$

3. $21 \div 7 = \underline{3}$

4. $30 \div 6 = \underline{5}$

5. $56 \div 8 = \underline{7}$

6. $45 \div 9 = \underline{5}$

7. done

8. $\times$

9. $+$

10. $-$

11.
$$\begin{array}{r} {}^1 1, {}^1 5\,8\,2 \\ +\,3,6\,2\,4 \\ \hline 5,2\,0\,6 \end{array}$$

12.
$$\begin{array}{r} 7,1\,3\,2 \\ +\,5,3\,3\,3 \\ \hline 12,4\,6\,5 \end{array}$$

13.
$$\begin{array}{r} {}^1 2, {}^1 8\,5\,2 \\ +\,4,2\,6\,3 \\ \hline 7,1\,1\,5 \end{array}$$

14. $22 \times 40 = 880$

15. $11 \times 60 = 660$

16. $12 \times 40 = 480$

17. $200 - 17 = 183$ pieces

18. $1,436 + 1,529 = 2,965$ mi

Systematic Review 15E

1. 107,873; one hundred seven thousand, eight hundred seventy-three

2. $3 \times 1,000,000,000,000 + 5 \times 100$

3. $56 \div 7 = \underline{8}$

4. $48 \div 6 = \underline{8}$

5. $30 \div 5 = \underline{6}$

6. $21 \div 3 = \underline{7}$

7. $\times$

8. $-$

9. $\div$

10. $+$

11.
$$\begin{array}{r} 1,{}^3 \cancel{4}\,{}^1 2\,6 \\ -\quad\;\;8\,7\,3 \\ \hline 5\,5\,3 \end{array}$$

12.
$$\begin{array}{r} {}^3 \cancel{4},2\,{}^7 \cancel{8}\,{}^1 3 \\ -\quad\;\;9\,5\,5 \\ \hline 3,3\,2\,8 \end{array}$$

13.
$$\begin{array}{r} {}^5\!6,{}^4\!2\,{}^1\!3\,4\ {}^1\!1 \\ -\quad 3\ 7\ 8 \\ \hline 5,\ 8\ \ 6\ 3 \end{array}$$

14. $14 \times 20 = 280$

15. $20 \times 100 = 2,000$

16. $45 \times 100 = 4,500$

17. $27 + 38 + 34 = 99$
$99 \div 3 = 33$ runs

18. $36 \times \$6 = \$216$
$36 \div 4 = 9$ gal

Systematic Review 15F

1. 8,472,600,000; eight billion,
four hundred seventy-two million,
six hundred thousand

2. $2 \times 1,000,000,000,000 +$
$2 \times 10,000,000,000 + 7 \times 1,000,000,000$

3. $63 \div 9 = \underline{7}$

4. $24 \div 8 = \underline{3}$

5. $42 \div 6 = \underline{7}$

6. $49 \div 7 = \underline{7}$

7. $\div$

8. $\div$

9. $+$

10. $-$

11.
$$\begin{array}{r} {}^1\!6,{}^1\!7\,3\,2 \\ 3,1\,5\,2 \\ +7,3\,2\,1 \\ \hline 1\,7,2\,0\,5 \end{array}$$

12.
$$\begin{array}{r} 5,9\,8\,9 \\ -\quad 6\,3\,2 \\ \hline 5,3\,5\,7 \end{array}$$

13.
$$\begin{array}{r} {}^1\!5,{}^1\!2\,3\,2 \\ 7,1\,1\,1 \\ +3,7\,6\,5 \\ \hline 1\,6,1\,0\,8 \end{array}$$

14. $11 \times 60 = 660$

15. $21 \times 40 = 840$

16. $40 \times 200 = 8,000$

17. $8 + 12 = 20$
$20 \div 2 = 10$
$10 \times 25 = 250$ sq ft

18. $2 \times 16 = 32$ ounces lost
32 quarters $\div 4 = \$8$

Lesson Practice 16A

1. done

2.
$$\begin{array}{r} 5\text{r. }3 \\ 4\overline{)23} \\ \underline{20} \\ 3 \end{array}$$

3.
$$\begin{array}{r} 8\text{r. }3 \\ 7\overline{)59} \\ \underline{56} \\ 3 \end{array}$$

4.
$$\begin{array}{r} 1\text{r. }6 \\ 7\overline{)13} \\ \underline{7} \\ 6 \end{array}$$

5.
$$\begin{array}{r} 4\text{r. }1 \\ 8\overline{)33} \\ \underline{32} \\ 1 \end{array}$$

6.
$$\begin{array}{r} 2\text{r. }5 \\ 8\overline{)21} \\ \underline{16} \\ 5 \end{array}$$

7.
$$\begin{array}{r} 9\text{r. }4 \\ 9\overline{)85} \\ \underline{81} \\ 4 \end{array}$$

8.
$$\begin{array}{r} 2\text{r. }2 \\ 9\overline{)20} \\ \underline{18} \\ 2 \end{array}$$

9.
$$\begin{array}{r} 7\text{r. }1 \\ 9\overline{)64} \\ \underline{63} \\ 1 \end{array}$$

10.
```
      6 r. 3
  6 ) 39
      36
       3
```

11.
```
      6 r. 4
  6 ) 40
      36
       4
```

12.
```
      2 r. 1
  6 ) 13
      12
       1
```

13.
```
      9 r. 1
  6 ) 46
      45
       1
```

14.
```
      3 r. 4
  5 ) 19
      15
       4
```

15.
```
      9 r. 3
  5 ) 48
      45
       3
```

16. 9 r. 3 ; 9 cookies each, 3 left over
```
      9 r. 3
  8 ) 75
      72
       3
```

17. 6 r. 1 ; 6 days, 1 left over
```
      6 r. 1
  2 ) 13
      12
       1
```

18. 9 r. 3 ; 9 dollars each, 3 bills left over
```
      9 r. 3
  5 ) 48
      45
       3
```

Lesson Practice 16B

1.
```
      2 r. 1
  3 ) 7
      6
      1
```

2.
```
      4 r. 2
  3 ) 14
      12
       2
```

3.
```
      9 r. 2
  3 ) 29
      27
       2
```

4.
```
      4 r. 1
  4 ) 17
      16
       1
```

5.
```
      4 r. 1
  2 ) 9
      8
      1
```

6.
```
      6 r. 1
  6 ) 37
      36
       1
```

7.
```
      4 r. 1
  7 ) 29
      28
       1
```

8.
```
      2 r. 3
  7 ) 17
      14
       3
```

9.
```
     10 r. 1
  5 ) 51
      50
       1
```

10.
```
      1 r. 2
  5 ) 7
      5
      2
```

11.
```
    8 r. 2
8 | 66
    64
     2
```

12.
```
    3 r. 3
8 | 27
    24
     3
```

13.
```
    8 r. 3
9 | 75
    72
     3
```

14.
```
    1 r. 5
9 | 14
     9
     5
```

15.
```
    3 r. 1
3 | 10
     9
     1
```

16. 5 r. 1; 5 cookies each, 1 left over
```
4 | 21
    20
     1
```

17. 4 r. 2; no, 2 extra
```
6 | 26
    24
     2
```

18. 4 r. 3; 4 teams, 3 extra
```
9 | 39
    36
     3
```

Lesson Practice 16C

1.
```
    9 r. 1
2 | 19
    18
     1
```

2.
```
    1 r. 1
2 | 3
    2
    1
```

3.
```
    3 r. 1
5 | 16
    15
     1
```

4.
```
    7 r. 2
5 | 37
    35
     2
```

5.
```
    8 r. 4
9 | 76
    72
     4
```

6.
```
    3 r. 2
9 | 29
    27
     2
```

7.
```
    9 r. 2
4 | 38
    36
     2
```

8.
```
    3 r.1
4 | 13
    12
     1
```

9.
```
    1 r. 2
7 | 9
    7
    2
```

10.
```
    3 r. 1
7 | 22
    21
     1
```

11.
```
    5 r. 1
3 | 16
    15
     1
```

12.
```
      6 r. 2
  3 ⌐ 20
       18
        2
```

13.
```
      2 r. 3
  8 ⌐ 19
       16
        3
```

14.
```
      9 r. 7
  8 ⌐ 79
       72
        7
```

15.
```
      8 r. 2
  6 ⌐ 50
       48
        2
```

16.
```
      7 r. 2 ; 7 weeks, 2 days
  7 ⌐ 51
       49
        2
```

17.
```
      6 r. 2 ; $6 apiece, $2 left over
  5 ⌐ 32
       30
        2
```

18.
```
      3 r. 2 ; 4 cars
  5 ⌐ 17
       15
        2
```

Systematic Review 16D

1.
```
      7 r. 2
  3 ⌐ 23
       21
        2
```

2.
```
     10 r. 1
  3 ⌐ 31
       30
        1
```

3.
```
      3 r. 2
  6 ⌐ 20
       18
        2
```

4.
```
      5 r. 1
  2 ⌐ 11
       10
        1
```

5.
```
      7 r. 2
  4 ⌐ 30
       28
        2
```

6.
```
      5 r. 5
  7 ⌐ 40
       35
        5
```

7.
```
      6 r. 3
  9 ⌐ 57
       54
        3
```

8.
```
      2 r. 2
  5 ⌐ 12
       10
        2
```

9.
```
      5 r. 3
  8 ⌐ 43
       40
        3
```

10.
```
    ¹  ¹
    6,718
  +2,452
    9,170
```

11.
```
    5,²3⁹⁰ ¹2
  −1, 2  3  8
    4, 0  6  4
```

12.
```
    4,⁵6⁰⁹ ¹2
  −3, 5  2  6
    1, 0  8  6
```

13. $11 \times 700 = 7,700$

14. $12 \times 300 = 3,600$

15. $20 \times 500 = 10,000$

16. 6,217,000,000

17.
```
      9 r. 1
  2 | 19
     18
      1
```
9 full loads, 1 bucket on last trip

18. $1{,}227 + 2{,}341 = 3{,}568$ animals

Systematic Review 16E

1.
```
      2r. 2
  4 | 10
      8
      2
```

2.
```
      6r. 3
  7 | 45
     42
      3
```

3.
```
      5r. 4
  9 | 49
     45
      4
```

4.
```
      7r. 4
  6 | 46
     42
      4
```

5.
```
      8r. 1
  3 | 25
     24
      1
```

6.
```
      6r. 2
  8 | 50
     48
      2
```

7.
```
      5r. 2
  5 | 27
     25
      2
```

8.
```
      2r. 1
  2 | 5
      4
      1
```

9.
```
      5r. 5
  6 | 35
     30
      5
```

10.
```
    ¹2,482
  + 7,902
   10,384
```

11.
```
    7,¹059
  + 3,470
   10,529
```

12.
```
    6,997
  - 2,961
    4,036
```

13. $11 \times 600 = 6{,}600$

14. $13 \times 200 = 2{,}600$

15. $22 \times 300 = 6{,}600$

16. $3 \times 1{,}000{,}000{,}000{,}000 + 4 \times 100{,}000{,}000{,}000 + 9 \times 10{,}000{,}000{,}000 + 1 \times 1{,}000{,}000{,}000$

17. 3r. 1; 3 CDs, $1 left over
```
  8 | 25
     24
      1
```

18. $20 + 13 + 8 + 7 = 48$
$48 \div 4 = 12$ lb

Systematic Review 16F

1.
```
      3r. 1
  2 | 7
      6
      1
```

2.
```
      4 r. 4
  5 | 24
     20
      4
```

3.
```
      7 r. 6
  8 | 62
     56
      6
```

4.
```
    9 r. 1
3 ) 28
    27
     1
```

5.
```
    9 r. 7
9 ) 88
    81
     7
```

6.
```
    8 r. 1
4 ) 33
    32
     1
```

7.
```
    9 r. 2
7 ) 65
    63
     2
```

8.
```
    1 r. 2
6 ) 8
    6
    2
```

9.
```
    7 r. 4
7 ) 53
    49
     4
```

10.
```
  4 5,¹0 ⁱ 1 42
 − 3, 9 7 1
   1, 1 7 1
```

11.
```
  9,0 ⁵6 ¹5
 −4,0 1 8
  5,0 4 7
```

12.
```
  ¹8,932
 +6,823
 15,755
```

13. $10 \times 700 = 7{,}000$

14. $21 \times 300 = 6{,}300$

15. $22 \times 400 = 8{,}800$

16. 7,349,000

17.
```
    8 r. 2 ; 8 presents, 2 bows left over
3 ) 26
    24
     2
```

18. rectangle:
$8 \times 10 = 80$ sq ft
trapezoid:
$5 + 7 = 12$
$12 \div 2 = 6$
$6 \times 6 = 36$ sq ft
$36 + 80 = 116$ sq ft total
$116 < 300$; yes

Lesson Practice 17A

1. done

2. done

3.
```
   33      30+3
 × 3    ×     3
   99      90+9
```

4.
```
     3          3
 ×33      ×30+3
    9         +9
   90        90
   99      90+9 = 99
```

5.
```
   431           431
 ×   2    ×        2
   862      800+60+2
```

6.
```
     2              2
 ×431      ×400+30+1
     2              2
    60             60
   800            800
   862      800+60+2 = 862
```

7. done

8. done

9.
```
     20
5 ) 100
  −100
     0
```

10.
```
    30
3 ) 90
  −90
    0
```

11.
```
     60
6 ) 360
  −360
     0
```

12.
```
      40
 4 )160
   -160
      0
```

Lesson Practice 17B

1.
```
   32        30+2
 ×  3       ×   3
   96        90+6
```

2.
```
    3           3
  ×32        ×30+2
    6           6
   90          90
   96        90+6 = 96
```

3.
```
   438      400+30+ 8
 ×   2     ×        2
   876      800+60+16
```

4.
```
     2              2
  ×438        ×400+30+8
    16             16
    60             60
   800            800
   876        800+70+6 = 876
```

5.
```
    13
   129      100+20+ 9
 ×   4     ×        4
   5 16     400+80+36
```

6.
```
     4              4
  ×129      ×100+20+ 9
    36            3 6
    80            8 0
   400           40 0
   516       400+80+36 = 516
```

7.
```
    20
 3 )60
   -60
     0
```

8.
```
     90
 9 )810
   -810
      0
```

9.
```
     30
 8 )240
   -240
      0
```

10.
```
    40
 2 )80
   -80
     0
```

11.
```
     60
 5 )300
   -300
      0
```

12.
```
     30
 7 )210
   -210
      0
```

Lesson Practice 17C

1.
```
   ²25        20+ 5
 ×   4      ×     4
   100        80+20
```

2.
```
     4            4
   ×25        ×20+5
    20           20
    80           80
   100        80+20 = 100
```

3.
```
   ¹¹22       100+ 2 0+ 2
 ×    5     ×           5
   6 10      500+100+10
```

4.
```
      5              5
  ×122      ×100+ 2 0+ 2
     10            1 0
    10 0          10 0
   5 0 0         500
   6 1 0      500+100+10 = 610
```

5.
```
   124      100+20+4
 ×   2     ×        2
   248      200+40+8
```

6.
$$\begin{array}{r} 2 \\ \times 124 \\ \hline 8 \\ 40 \\ 200 \\ \hline 248 \end{array}$$
$$\begin{array}{r} 2 \\ \times 100+20+4 \\ \hline 8 \\ 40 \\ 200 \\ \hline 200+40+8=248 \end{array}$$

7.
$$\begin{array}{r} 20 \\ 4\overline{)80} \\ -80 \\ \hline 0 \end{array}$$

8.
$$\begin{array}{r} 70 \\ 6\overline{)420} \\ -420 \\ \hline 0 \end{array}$$

9.
$$\begin{array}{r} 30 \\ 9\overline{)270} \\ -270 \\ \hline 0 \end{array}$$

10.
$$\begin{array}{r} 40 \\ 3\overline{)120} \\ -120 \\ \hline 0 \end{array}$$

11.
$$\begin{array}{r} 80 \\ 8\overline{)640} \\ -640 \\ \hline 0 \end{array}$$

12.
$$\begin{array}{r} 70 \\ 5\overline{)350} \\ -350 \\ \hline 0 \end{array}$$

Systematic Review 17D

1.
$$\begin{array}{r} 4^17 \\ \times\ \ 2 \\ \hline 834 \end{array}$$
$$\begin{array}{r} 400+10+\ 7 \\ \times\ \ \ \ \ \ \ \ \ \ 2 \\ \hline 800+20+14 \end{array}$$

2.
$$\begin{array}{r} 2 \\ \times 417 \\ \hline 14 \\ 20 \\ 800 \\ \hline 834 \end{array}$$
$$\begin{array}{r} 2 \\ \times 400+10+\ 7 \\ \hline 1\ 4 \\ 2\ 0 \\ 80\ 0 \\ \hline 800+20+14=834 \end{array}$$

3.
$$\begin{array}{r} 40 \\ 7\overline{)280} \\ -280 \\ \hline 0 \end{array}$$

4.
$$\begin{array}{r} 30 \\ 6\overline{)180} \\ -180 \\ \hline 0 \end{array}$$

5.
$$\begin{array}{r} 4r.\ 2 \\ 9\overline{)38} \\ -36 \\ \hline 2 \end{array}$$

6.
$$\begin{array}{r} 8r.\ 3 \\ 8\overline{)67} \\ -64 \\ \hline 3 \end{array}$$

7. $8+2=10$
$10\div2=5$
$5\times5=25$ sq ft

8. $2\times7=14$
$14\div2=7$ sq in

9. $25\times25=625$ sq mi

10. $5\times16=80$

11. $20\div4=5$

12. $12\div4=3$

13. done

14. $3\times2,000=6,000$

15. $7\times2,000=14,000$

16.
$$\begin{array}{r} 5r.\ 1;\ 5\ pieces,\ 1\ foot\ long \\ 3\overline{)16} \\ 15 \\ \hline 1 \end{array}$$

17. $312\times3=936$ miles

18. $2\times16=32$ oz
$32>30$
The 2-lb can is more.

Systematic Review 17E

1.
$$\begin{array}{r} 432 \\ \times\ \ 5 \\ \hline 2,160 \end{array}$$
$$\begin{array}{r} 400+3\ 0+2 \\ \times\ \ \ \ \ \ \ \ \ \ 5 \\ \hline 2,000+150+10 \end{array}$$

2.

```
        5                5
      ×432    ×400+ 3 0+ 2
       10               1 0
      150              15 0
     2000            2 00 0
     2,160    2,000+150+10 = 2,160
```

3.

```
        90
    5 ⟌450
      −450
         0
```

4.

```
        70
    8 ⟌560
      −560
         0
```

5.

```
       3r. 1
    4 ⟌13
       12
        1
```

6.

```
       6r. 2
    6 ⟌38
      −36
        2
```

7. $1+7 = 8$

$8 \div 2 = 4$

$4 \times 6 = 24$ sq ft

8. $3 \times 4 = 12$

$12 \div 2 = 6$ sq in

9. $7 \times 5 = 35$ sq in

10. $8 \times 16 = 128$

11. $18 \div 3 = 6$

12. $24 \times 2 = 48$

13. $5 \times 2,000 = 10,000$

14. $9 \times 2,000 = 18,000$

15. $4 \times 2,000 = 8,000$

16. 7r. 2; 7 times, 2 minutes to spare

```
    4 ⟌30
       28
        2
```

17. $\$112 \times 4 = \$448$

18. $5 \times 1,000,000 + 6 \times 100,000 +$

Systematic Review 17F

1.

```
      341     300+ 4 0+1
    ×   3   ×         3
    1,023    900+120+3
```

2.

```
        3                3
      ×341    ×300+ 4 0+1
        3                3
      120              120
      900              900
    1,023    900+120+3 = 1,023
```

3.

```
        30
    2 ⟌60
       60
        0
```

4.

```
        80
    9 ⟌720
      −720
        0
```

5.

```
       6r. 3
    7 ⟌45
       42
        3
```

6.

```
       9r. 3
    4 ⟌39
       36
        3
```

7. $3+9 = 12$

$12 \div 2 = 6$

$6 \times 4 = 24$ sq ft

8. $6 \times 8 = 48$

$48 \div 2 = 24$ sq in

9. $30 \times 20 = 600$ sq in

10. $7 \times 16 = 112$

11. $6 \times 4 = 24$

12. $21 \div 3 = 7$

13. $2 \times 2,000 = 4,000$

14. $8 \times 2,000 = 16,000$

15. $6 \times 2,000 = 12,000$

16. $75 \div 9 = 8$ r. 3; 8 full loads + 3 in last load = total of 9 loads

17. $231 \times 3 = 693$ jelly beans

18. $6 \times 1,000,000,000,000 + 2 \times 100,000,000,000$

Lesson Practice 18A

1. done
2. done

3.
$$\begin{array}{r} 20 \\ 3\overline{)60} \\ \underline{60} \\ 0 \end{array} \qquad \begin{array}{r} 3 \\ \times 20 \\ \hline 60 \end{array}$$

4.
$$\begin{array}{r} 3 \\ 20 \\ 2\overline{)46} \\ \underline{-40} \\ 6 \\ \underline{-6} \\ 0 \end{array} \qquad \begin{array}{r} 2 \\ \times 20+3 \\ 6 \\ \underline{40} \\ 40+6=46 \end{array}$$

5.
$$\begin{array}{r} 2r.\,1 \\ 20 \\ 3\overline{)67} \\ \underline{-60} \\ 7 \\ \underline{-6} \\ 1 \end{array} \qquad \begin{array}{r} 3 \\ \times 20+2 \\ 6 \\ \underline{60} \\ 60+6\;+1=67 \end{array}$$

6.
$$\begin{array}{r} 20 \\ 2\overline{)40} \\ \underline{-40} \\ 0 \end{array} \qquad \begin{array}{r} 2 \\ \times 20 \\ \hline 40 \end{array}$$

7.
$$\begin{array}{r} 2 \\ 30 \\ 3\overline{)96} \\ \underline{-90} \\ 6 \\ \underline{-6} \\ 0 \end{array} \qquad \begin{array}{r} 3 \\ \times 30+2 \\ 6 \\ \underline{90} \\ 90+6=96 \end{array}$$

8.
$$\begin{array}{r} 1r.\,1 \\ 40 \\ 2\overline{)83} \\ \underline{-80} \\ 3 \\ \underline{-2} \\ 1 \end{array} \qquad \begin{array}{r} 2 \\ \times 40+1 \\ 2 \\ \underline{80} \\ 80+2\;+1=83 \end{array}$$

9.
$$\begin{array}{r} 4 \\ 30 \\ 2\overline{)68} \\ \underline{-60} \\ 8 \\ \underline{-8} \\ 0 \end{array} \qquad \begin{array}{r} 2 \\ \times 30+4 \\ 8 \\ \underline{60} \\ 60+8=68 \end{array}$$

34 pages

10.
$$\begin{array}{r} 1 \\ 10 \\ 5\overline{)55} \\ \underline{-50} \\ 5 \\ \underline{-5} \\ 0 \end{array} \qquad \begin{array}{r} 5 \\ \times 10+1 \\ 5 \\ \underline{50} \\ 50+5=55 \end{array}$$

11 feet

Lesson Practice 18B

1.
$$\begin{array}{r} 3 \\ 20 \\ 2\overline{)46} \\ \underline{-40} \\ 6 \\ \underline{-6} \\ 0 \end{array} \qquad \begin{array}{r} 2 \\ \times 20+3 \\ 6 \\ \underline{40} \\ 40+6=46 \end{array}$$

2.
$$\begin{array}{r} 1\,r.\,3 \\ 20 \\ 4\overline{)87} \\ \underline{-80} \\ 7 \\ \underline{-4} \\ 3 \end{array} \qquad \begin{array}{r} 4 \\ \times 20+1 \\ 4 \\ \underline{80} \\ 80+4\;+3=87 \end{array}$$

3.
$$\begin{array}{r} 1\,r.2 \\ 20 \\ 3\overline{)65} \\ \underline{-60} \\ 5 \\ \underline{-3} \\ 2 \end{array} \qquad \begin{array}{r} 3 \\ \times 20+1 \\ 3 \\ \underline{60} \\ 60+3\;+2=65 \end{array}$$

4.

```
        1          2
       10      ×10+1
    2⟌22          2
     -20         20
       2     20 + 2 = 22
      -2
       0
```

5.

```
        2          2
       30      ×30+2
    2⟌64          4
     -60         60
       4     60 + 4 = 64
      -4
       0
```

6.

```
      30 r.1     3
    3⟌ 91      ×30
      -90     90 + 1 = 91
        1
```

7.

```
      1 r. 1       4
       10      ×10+1
    4⟌ 45          4
      -40         40
        5   40 + 4 + 1 = 45
       -4
        1
```

8.

```
        4          2
       40      ×40+4
    2⟌ 88          8
      -80         80
        8    80 + 8 = 88
       -8
        0
```

9.

```
        1          6
       10      ×10+1
    6⟌ 66          6
      -60         60
        6    60 + 6 = 66
       -6
        0
```

11 jelly beans

10.

```
      1 r. 2       3
       10      ×10+1
    3⟌ 35          3
      -30         30
        5   30 + 3 + 2 = 35
       -3
        2
```

11 times; 2 tokens left over

Lesson Practice 18C

1.

```
        1          7
       10      ×10+1
    7⟌ 77          7
      -70         70
        7    70 + 7 = 77
       -7
        0
```

2.

```
      1 r. 3       5
       10      ×10+1
    5⟌ 58          5
      -50         50
        8   50 + 5 + 3 = 58
       -5
        3
```

3.

```
      2 r. 1       2
       10      ×10+2
    2⟌ 25          4
      -20         20
        5   20 + 4 + 1 = 25
       -4
        1
```

4.

```
      2 r. 1       3
       10      ×10+2
    3⟌ 37          6
      -30         30
        7   30 + 6 + 1 = 37
       -6
        1
```

5.

```
     20 r. 1       4
    4⟌ 81       ×20
      -80     80 + 1 = 81
```

6.
```
    30        2
  2│60      ×30
   -60       60
     0
```

7.
```
   10 r. 7     8
  8│87       ×10
   -80      80 + 7 = 87
     7
```

8.
```
    2 r. 2       3
    30        ×30+2
  3│98          6
   -90          90
     8      90+6 +2 = 98
    -6
     2
```

9.   2 ; 12 gallons
```
    10
  4│48
   -40
     8
    -8
     0
```

10.   2 r. 3 ; 12 books, $3 left
```
    10
  8│99
   -80
    19
   -16
     3
```

Systematic Review 18D

1.
```
    1 r. 1        3
    10        ×10+1
  3│34           3
   -30           30
     4      30+3 + 1 = 34
    -3
     1
```

2.
```
    2 r. 1        4
    20        ×20+2
  4│89           8
   -80           80
     9      80+8 + 1 = 89
    -8
     1
```

3.
```
   10 r. 4       6
  6│64         ×10
   -60      60 + 4 = 64
     4
```

4.
```
    7 r. 1       8
  8│57         × 7
   -56      56  + 1 = 57
     1
```

5.
```
    9 r. 3       4
  4│39         × 9
   -36      36  + 3 = 39
     3
```

6.
```
    40          7
  7│280       ×40
   -280       280
     0
```

7. $16 + 25 + 48 = \underline{89}$
8. $423 + 32 + 7 = \underline{462}$
9. $8 + 2 + 1 + 9 = \underline{20}$
10. $2 \times 2,000 = 4,000$
11. $10 \times 2,000 = 20,000$
12. $7 \times 2,000 = 14,000$
13. done
14. $5,280 \times 2 = 10,560$
15. $5,280 \times 8 = 42,240$
16. $88 \div 8 = 11$ pieces
17. $4 \times 5,280 = 21,120$ ft
18. $2,465 + 2,153 = 4,618$ mi

Systematic Review 18E

1.
```
        4          2
       10        ×10+4
    2│28           8
     −20          20
       8        20+8 = 28
      −8
       0
```

2.
```
      20 r. 1      3
    3│61         ×20
     −60        60 + 1= 61
       1
```

3.
```
       1 r. 2      7
      10        ×10+1
    7│79           7
     −70          70
       9       70+7 + 2 = 79
      −7
       2
```

4.
```
       2 r. 1      5
    5│11         × 2
     −10        10 + 1= 11
       1
```

5.
```
       9 r. 3      9
    9│84         × 9
     −81        81 + 3 = 84
       3
```

6.
```
       40          6
    6│240        ×40
     −240        240
       0
```

7. $6+4+3+2+9 = \underline{24}$

8. $167+4+58 = \underline{229}$

9. $11+15+41 = \underline{67}$

10. $4 \times 2,000 = 8,000$

11. $36 \times 3 = 108$

12. $18 \times 2 = 36$

13. $5 \times 5,280 = 26,400$

14. $10 \times 5,280 = 52,800$

15. $7 \times 5,280 = 36,960$

16. $41 \div 4 = 10$ r. 1; 10 blocks, 1 left over

17. $72 \div 6 = 12$ trips
6 tons $\times 2,000 = 12,000$ pounds

18. $2+2+0+8+4+3+2 = 21$
$21 \div 7 = 3$ children

Systematic Review 18F

1.
```
      10 r. 4      5
    5│54         ×10
     −50       50 + 4 = 54
       4
```

2.
```
      10 r. 1      8
    8│81         ×10
     −80       80 + 1 = 81
       1
```

3.
```
        2          2
       40        ×40+2
    2│84           4
     −80          80
       4       80  +4 = 84
      −4
       0
```

4.
```
       5 r. 1      3
    3│16         × 5
     −15       15  +1= 16
       1
```

5.
```
       5 r. 3      4
    4│23         × 5
     −20       20  +3 = 23
       3
```

6.
```
       80          9
    9│720        ×80
     −720        720
       0
```

7. $245+961+102 = \underline{1,308}$

8. $25+631+ 40 = \underline{696}$

9. $9+8+6+2+4 = \underline{29}$

10. $3 \times 2,000 = 6,000$

11. $76 \times 4 = 304$

12. $11 \times 16 = 176$

13. $4 \times 5,280 = 21,120$

14. $9 \times 5,280 = 47,520$

15. $6 \times 5,280 = 31,680$

16. $15 \div 4 = 3$ r. 3; 3 batches, 3 cups left over

17. $12 + 8 + 16 = 36$

    $36 \div 3 = 12$ hours

18. $\$55 + \$38 = \$93$

    $\$150 - \$93 = \$57$

Lesson Practice 19A

1. done

2. done

3.
```
        3          3
       10        ×213
      200           9
   3√639           30
      600          600
       39          639
       30
        9
        9
        0
```

4.
```
        3          2
       40        ×343
    300r. 1         6
   2√687           80
      600          600
       87        686 +1= 687
       80
        7
        6
        1
```

5.
```
        1          4
     40r. 2       ×41
   4√166           4
      160          160
        6        164 +2=166
        4
        2
```

6.
```
        6          2
       00        ×306
     300r. 1        12
   2√613            00
      600          600
       13        612 +1= 613
       12
        1
```

7.
```
        2          6
       10        ×112
    100r. 4        12
   6√676           60
      600          600
       76        672 +4 = 676
       60
       16
       12
        4
```
112 pennies; 4 pennies left over

8.
```
        1          4
       10        ×211
      200           4
   4√844           40
      800          800
       44          844
       40
        4
        4
        0
```
211 antelope

Lesson Practice 19B

1.
```
        5          8
     50r. 2       ×55
   8√442           40
      400          400
       42        440 +2 = 442
       40
        2
```

2.
```
        3           3
       10         ×113
      100            9
   3│339            30
      300           300
       39           339
       30
        9
        9
        0
```

3.
```
        9           3
       30         ×139
    100r. 2         27
   3│419            90
      300           300
      119         417 +2 = 419
       90
       29
       27
        2
```

4.
```
        3           2
       20         ×323
      300            6
   2│646            40
      600           600
       46           646
       40
        6
        6
        0
```

5.
```
        7           4
       20         ×227
      200           28
   4│908            80
      800           800
      108           908
       80
       28
       28
        0
```

6.
```
        1           5
     30r.1         ×31
   5│156             5
      150           150
        6         155 +1 = 156
        5
        1
```

7. $333 \div 3 = 111$ yards

8. $669 \div 6 = 111$ r.3; 112 groups in all, 3 people in last group

Lesson Practice 19C

1.
```
        5           3
       00         ×105
    100 r. 2        15
   3│317            00
      300           300
       17         315 + 2 = 317
       15
        2
```

2.
```
        7           7
       10         ×117
    100 r. 5        49
   7│824            70
      700           700
      124         819 + 5 = 824
       70
       54
       49
        5
```

3.
```
        1           4
       10         ×111
    100 r. 1         4
   4│445            40
      400           400
       45         444 + 1 = 445
       40
        5
        4
        1
```

4.

```
       6        4
     60 r. 2   ×66
  4│266        24
    240        240
     26        264 + 2 = 266
     24
      2
```

5.

```
        0        6
       10      ×110
      100 r. 5    0
  6│665          60
    600          600
     65          660 + 5 = 665
     60
      5
      0
      5
```

6.

```
       1        9
      90      ×91
  9│819         9
    810        810
      9        819
      9
      0
```

7. 802 ÷ 2 = 401 jugs

8. 550 ÷ 5 = 110 cages

Systematic Review 19D

1.

```
       8
      00
     100 r. 3
  4│435
    400
     35
     32
      3
```

2.

```
       4
    ×108
      32
      00
     400
    432 + 3 = 435
```

3.

```
       4
      40 r. 2
  6│266
    240
     26
     24
      2
```

4.

```
       6
    ×44
     24
    240
    264 + 2 = 266
```

5.

```
     1 1 1
     2,962
    +3,148
     6,110
```

6.

```
     1,¹250
    +  351
     1,601
```

7.

```
     ³3,¹076
    +1,942
     5,018
```

8. 6 × 2,000 = 12,000

9. 2 × 5,280 = 10,560

10. 8 × 5,280 = 42,240

11. parallel

12. 6 × 1,000,000,000,000 + 5 × 100,000,000,000 + 8 × 10,000,000,000 + 3 × 1,000,000,000

13. 6,000,000,705

14. 7,821 − 4,568 = 3,253 miles

15. 69 ÷ 6 = 11 r. 3; 11 teams, 3 left

16. 3 × 2,000 = 6,000 lb

Systematic Review 19E

1.

```
      20
     300
  3│960
    900
     60
     60
      0
```

2.
```
      3
   ×320
      0
     60
    900
    960
```

3.
```
      7
     50 r. 2
  5│287
    250
     37
     35
      2
```

4.
```
      5
    ×57
     35
    250
    285 + 2 = 287
```

5.
```
   7  2
  5,8 3 2
  -2, 1 7 4
   3, 6 5 8
```

6.
```
  8   5
  9,0 6 7
  -   1 5 8
   8, 9 0 9
```

7.
```
     5
  4,9 6 5
  -3,4 5 6
   1,5 0 9
```

8. 20 ÷ 4 = 5

9. 32 ÷ 4 = 8

10. 16 ÷ 2 = 8

11. true

12. 3,400,000,000,800

13. 639 ÷ 3 = 213 yd

14. 3 × 5,280 = 15,840 ft

15. Kimberly: 560 + 237 = 797
850 − 797 = 53 words

16. 3 × 2,000 = 6,000
6,000 > 3,500
3 tons is better.

Systematic Review 19F

1.
```
      4
     00
    200 r. 1
  4│817
    800
     17
     16
      1
```

2.
```
      4
   ×204
     16
     00
    800
    816 + 1 = 817
```

3.
```
      1
     00
    200
  2│402
    400
      2
      2
      0
```

4.
```
      2
   ×201
      2
     00
    400
    402
```

5. done

6.
```
    46 1
   × 85
      3
   ¹2 0 0 5
      4
    3 2 8 8
   3 9,18 5
```

7.
```
    5 5 8
   × 39
   ¹4 7
   ¹4 5 5 2
    1 2
   ¹1 5 5 4
   21,7 6 2
```

8. $100 \times 3 = 300$

9. $4 \times 5,280 = 21,120$

10. $17 \times 16 = 272$

11. perpendicular

12. $700 \div 7 = 100$ weeks

13. $1,251,621$

14. $131 + 155 + 100 + 102 = 488$
    $488 \div 4 = \$122$

15. $11 \times 10 = 110$; $110 - 17 = 93$ jars

16.
$$
\begin{array}{r}
{}^1{}^1 1,8\,{}^5 9,{}^7 4\,8 \\
+\ \ 9,2\,3\,0,6\,8\,7 \\
\hline
2\,1,0\,9\,0,4\,3\,5
\end{array}
$$

Lesson Practice 20A

1. done

2. done

3.
$$
\begin{array}{r}
50\frac{4}{7} \\
7\overline{)354} \\
\underline{350} \\
4
\end{array}
\qquad
\begin{array}{r}
7 \\
\times\ 50 \\
\hline
350 \\
r.4 \\
\hline
354
\end{array}
$$

4.
$$
\begin{array}{r}
100\frac{5}{9} \\
9\overline{)905} \\
\underline{900} \\
5
\end{array}
\qquad
\begin{array}{r}
9 \\
\times 100 \\
\hline
900 \\
r.5 \\
\hline
905
\end{array}
$$

5.
$$
\begin{array}{r}
70 \\
6\overline{)420} \\
\underline{420} \\
0
\end{array}
\qquad
\begin{array}{r}
6 \\
\times 70 \\
\hline
420
\end{array}
$$

6.
$$
\begin{array}{r}
121\frac{6}{8} \\
8\overline{)974} \\
\underline{800} \\
174 \\
\underline{160} \\
14 \\
\underline{8} \\
6
\end{array}
\qquad
\begin{array}{r}
8 \\
\times 121 \\
\hline
8 \\
160 \\
\underline{800} \\
968 \\
r.6 \\
\hline
974
\end{array}
$$

7. $\$435 \div 2 = \$217\frac{1}{2}$

8. $185 \div 3 = 61\frac{2}{3}$ yd

Lesson Practice 20B

1.
$$
\begin{array}{r}
41\frac{4}{6} \\
6\overline{)250} \\
\underline{240} \\
10 \\
\underline{6} \\
4
\end{array}
\qquad
\begin{array}{r}
6 \\
\times 41 \\
\hline
6 \\
240 \\
246 \\
r.4 \\
\hline
250
\end{array}
$$

2.
$$
\begin{array}{r}
34\frac{6}{8} \\
8\overline{)278} \\
\underline{240} \\
38 \\
\underline{32} \\
6
\end{array}
\qquad
\begin{array}{r}
8 \\
\times 34 \\
\hline
32 \\
240 \\
272 \\
r.6 \\
\hline
278
\end{array}
$$

3.
$$
\begin{array}{r}
53\frac{2}{4} \\
4\overline{)214} \\
\underline{200} \\
14 \\
\underline{12} \\
2
\end{array}
\qquad
\begin{array}{r}
4 \\
\times 53 \\
\hline
12 \\
200 \\
212 \\
r.2 \\
\hline
214
\end{array}
$$

4.
$$
\begin{array}{r}
116\frac{5}{7} \\
7\overline{)817} \\
\underline{700} \\
117 \\
\underline{70} \\
47 \\
\underline{42} \\
5
\end{array}
\qquad
\begin{array}{r}
7 \\
\times 116 \\
\hline
42 \\
70 \\
700 \\
812 \\
r.5 \\
\hline
817
\end{array}
$$

5.
$$
\begin{array}{r}
109\frac{4}{6} \\
6\overline{)658} \\
\underline{600} \\
58 \\
\underline{54} \\
4
\end{array}
\qquad
\begin{array}{r}
6 \\
\times 109 \\
\hline
54 \\
00 \\
600 \\
654 \\
r.4 \\
\hline
658
\end{array}
$$

6.
$$
\begin{array}{r}
12\frac{7}{9} \\
9\overline{)115} \\
\underline{90} \\
25 \\
\underline{18} \\
7
\end{array}
\qquad
\begin{array}{r}
9 \\
\times 12 \\
\hline
18 \\
90 \\
108 \\
r.7 \\
\hline
115
\end{array}
$$

7. $\$145 \div \$7 = 20\frac{5}{7}$ weeks

8. $253 \div 2 = 126\frac{1}{2}$ days

Lesson Practice 20C

1.
```
    95 2/8        8
8)762          ×95
  720           40
   42          720
   40          760
    2          r.2
               762
```

2.
```
    52 3/6        6
6)315          ×52
  300           12
   15          300
   12          312
    3          r.3
               315
```

3.
```
    33 5/7        7
7)236          ×33
  210           21
   26          210
   21          231
    5          r.5
               236
```

4.
```
    27 3/5        5
5)138          ×27
  100           35
   38          100
   35          135
    3          r.3
               138
```

5.
```
   170 1/2        2
2)341          ×170
  200            0
  141          140
  140          200
    1          340
               r.1
               341
```

6.
```
   34 2/3         3
3)104          ×34
   90           12
   14           90
   12          102
    2          r.2
               104
```

7. $\$124 \div 8 = \$15\frac{4}{8}$

8. $275 \div 4 = 68\frac{3}{4}$ pails

Systematic Review 20D

1.
```
   141 1/6
6)847
  600
  247
  240
    7
    6
    1
```

2.
```
      6
   ×141
      6
    240
    600
    846
    r.1
    847
```

3.
```
   20 4/5
5)104
  100
    4
```

4.
```
     5
   ×20
   100
   r.4
   104
```

5. done

6. $16 \div 4 = 4$ ft

7. $14 \div 7 = 2$ in

8. $925 \div 2 = 462\frac{1}{2}$ miles

9. $15 \div 6 = 2\frac{3}{6}$ pieces
10. $8 \times 5,280 = 42,240$ ft
11. perpendicular
12. $35 \times 2,000 = 70,000$ lb
13. $63 + 79 + 72 = 214$

   $214 \div 3 = 71\frac{1}{3}$ degrees
14. $41 + 52 + 60 = 153$

   $153 \div 3 = 51$ degrees
15. $1,912 - 1,787 = 125$ years

Systematic Review 20E

1.
$$8\overline{)169} \quad 21\frac{1}{8}$$
$$\underline{160}$$
$$9$$
$$\underline{8}$$
$$1$$

2.
$$\begin{array}{r} 8 \\ \times 21 \\ \hline 8 \\ 160 \\ \hline 168 \\ r.1 \\ \hline 169 \end{array}$$

3.
$$7\overline{)724} \quad 103\frac{3}{7}$$
$$\underline{700}$$
$$24$$
$$\underline{21}$$
$$3$$

4.
$$\begin{array}{r} 7 \\ \times 103 \\ \hline 21 \\ 00 \\ 700 \\ \hline 721 \\ r.3 \\ \hline 724 \end{array}$$

5. $50 \div 5 = 10$ in
6. $15 \div 5 = 3$ ft
7. $72 \div 9 = 8$ in
8. $45 \div 9 = 5$ minutes

9. $67 \div 5 = 13$ r. 2; 13 packages, 2 stamps left over
10. $\$350 \div 8 = \$43\frac{6}{8}$
11. $2,000 \times 16 = 32,000$ ounces
12. $486 \div 4 = 121\frac{2}{4}$ dollars
13. yes
14. $1,959 - 1,788 = 171$ years
15. $60 \times 24 = 1,440$ minutes

Systematic Review 20F

1.
$$4\overline{)725} \quad 181\frac{1}{4}$$
$$\underline{400}$$
$$325$$
$$\underline{320}$$
$$5$$
$$\underline{4}$$
$$1$$

2.
$$\begin{array}{r} 4 \\ \times 181 \\ \hline 4 \\ 320 \\ 400 \\ \hline 724 \\ r.1 \\ \hline 725 \end{array}$$

3.
$$9\overline{)467} \quad 51\frac{8}{9}$$
$$\underline{450}$$
$$17$$
$$\underline{9}$$
$$8$$

4.
$$\begin{array}{r} 9 \\ \times 51 \\ \hline 9 \\ 450 \\ \hline 459 \\ r.8 \\ \hline 467 \end{array}$$

5. $100 \div 9 = 11\frac{1}{9}$ in
6. $25 \div 6 = 4\frac{1}{6}$ ft

7. $59 \div 8 = 7\frac{3}{8}$ in

8. $528 \div 3 = 176$ lb

9. $444 \div 8 = 55\frac{4}{8}$ miles

10. $2 \times 5{,}280 = 10{,}560$ ft

    $10{,}560 > 10{,}000$; Bethany ran farther.

11. $\$315 \div 3 = \$105$

    $\$112 > \$105$; $\$315$ for 3 tons is cheaper.

12. $10 \times 10 = 100$ sq mi

13. 4

14. $2 \times 70 = 140$

    $160 - 140 = 20$ lb

15. $20 + 15 = 35$ lb

Lesson Practice 21A

1. 50

2. 20

3. 50

4. 600

5. 200

6. 900

7. 2,000

8. 3,000

9. 6,000

10. done

11. done

12. $$\begin{array}{r} 200 \\ 2\overline{)500} \end{array}$$

13. $$\begin{array}{r} 244\frac{1}{2} \\ 2\overline{)489} \\ \underline{400} \\ 89 \\ \underline{80} \\ 9 \\ \underline{8} \\ 1 \end{array}$$

14. $$\begin{array}{r} 100 \\ 3\overline{)400} \end{array}$$

15. $$\begin{array}{r} 118\frac{2}{3} \\ 3\overline{)356} \\ 300 \\ 56 \\ \underline{30} \\ 26 \\ \underline{24} \\ 2 \end{array}$$

16. $400 \div 3 \approx 100$

    $395 \div 3 = 131\frac{2}{3}$ yd

Lesson Practice 21B

1. 30

2. 70

3. 90

4. 400

5. 600

6. 700

7. 2,000

8. 3,000

9. 3,000

10. $$\begin{array}{r} 300 \\ 2\overline{)700} \end{array}$$

11. $$\begin{array}{r} 367\frac{1}{2} \\ 2\overline{)735} \\ 600 \\ 135 \\ 120 \\ 15 \\ \underline{14} \\ 1 \end{array}$$

12. $500 \div 6 \approx 80$

13. $487 \div 6 = 81\frac{1}{6}$

14. $900 \div 9 = 100$

15. $921 \div 9 = 102\frac{3}{9}$

16. $900 \div 8 \approx 100$

    $938 \div 8 = 117\frac{2}{8}$ gallons

Lesson Practice 21C

1. 20
2. 40
3. 60
4. 500
5. 700
6. 200
7. 2,000
8. 2,000
9. 9,000
10. $$3\overline{)400} \quad \frac{100}{}$$
11.
$$3\overline{)436} \quad 145\frac{1}{3}$$
$$\underline{300}$$
$$136$$
$$\underline{120}$$
$$16$$
$$\underline{15}$$
$$1$$
12. $800 \div 2 = 400$
13. $751 \div 2 = 375\frac{1}{2}$
14. $800 \div 5 \approx 100$
15. $845 \div 5 = 169$
16. $800 \div 4 = 200$
    $820 \div 4 = 205$ jugs

Systematic Review 21D

1. 80
2. 200
3. 4,000
4. $900 \div 3 = 300$
5. $912 \div 3 = 304$
6. $378 \div 7 = 54$
7. $7 \times 54 = 378$
8. $559 \div 9 = 62\frac{1}{9}$
9. $9 \times 62 = 558$
   $558 + 1 = 559$
10. $7 \times 100,000 + 5 \times 10,000$
11. $3 \times 1,000,000 + 4 \times 100,000$
12. $8 \times 1,000,000,000$

13. $24 \div 3 = 8$ baskets
14. $11 \times 5,280 = 58,080$ ft
15. $3 \times 5 = 15$ cards
16. $\$375 \div 4 = \$93\frac{3}{4}$

Systematic Review 21E

1. 50
2. 200
3. 7,000
4. $800 \div 5 \approx 100$
5. $776 \div 5 = 155\frac{1}{5}$
6. $225 \div 4 = 56\frac{1}{4}$
7. $4 \times 56 = 224$
   $224 + 1 = 225$
8. $628 \div 6 = 104\frac{4}{6}$
9. $6 \times 104 = 624$
   $624 + 4 = 628$
10. $365 + 498 = 863$
11. $863 - 125 = 738$
12. $671 \times 32 = 21,472$
13. $21 \div 7 = 3$ per day
14. $10 \times 2,000 = 20,000$ lb
15. $70 \div 7 = 10$ yards
16. $7 \times 3 = 21$ feet wide
    $10 \times 3 = 30$ feet long

Systematic Review 21F

1. 20
2. 100
3. 8,000
4. $1,000 \div 7 \approx 100$
5. $980 \div 7 = 140$
6. $463 \div 3 = 154\frac{1}{3}$
7. $3 \times 154 = 462$
   $462 + 1 = 463$
8. $336 \div 8 = 42$
9. $8 \times 42 = 336$
10. $1,345 + 601 = 1,946$
11. $4,532 - 2,055 = 2,477$

12. $2{,}891 \times 12 = 34{,}692$

13. $22 + 26 = 48$
$48 \div 2 = 24$
$24 \times 19 = 456$ sq in

14. $58 + 67 + 91 + 88 = 304$
$304 \div 4 = 76$ pods

15. $484 \div 2 = 242$ quart jars

16. $15 \times 18 = 270$
$270 \div 2 = 135$ sq ft

Lesson Practice 22A

1. done

2.
$$13 \overline{)968} \quad 74\tfrac{6}{13}$$
910
58
52
6
$74 \times 13 = 962$
$962 + 6 = 968$

3.
$$12 \overline{)785} \quad 65\tfrac{5}{12}$$
720
65
60
5
$65 \times 12 = 780$
$780 + 5 = 785$

4.
$$21 \overline{)483} \quad 23$$
420
63
63
0
$23 \times 21 = 483$

5.
$$11 \overline{)638} \quad 58$$
550
88
88
0
$58 \times 11 = 638$

6.
$$15 \overline{)377} \quad 25\tfrac{2}{15}$$
300
77
75
2
$25 \times 15 = 375$
$375 + 2 = 377$

7. $450 \div 18 = 25$ people

8. $390 \div 26 = 15$ days

Lesson Practice 22B

1.
$$13 \overline{)169} \quad 13$$
130
39
39
0
$13 \times 13 = 169$

2.
$$41 \overline{)506} \quad 12\tfrac{14}{41}$$
410
96
82
14
$12 \times 41 = 492$
$492 + 14 = 506$

3.
$$61 \overline{)728} \quad 11\tfrac{57}{61}$$
610
118
61
57
$11 \times 61 = 671$
$671 + 57 = 728$

4.
```
        41
  15 ) 615
       600
        15
        15
         0
```
$41 \times 15 = 615$

5.
```
        24  20/26
  26 ) 644
       520
       124
       104
        20
```
$24 \times 26 = 624$
$624 + 20 = 644$

6.
```
        11  6/75
  75 ) 831
       750
        81
        75
         6
```
$11 \times 75 = 825$
$825 + 6 = 831$

7. $315 \div 15 = 21$ bags

8. $\$732 \div 12 = \$61$

Lesson Practice 22C

1.
```
        22
  12 ) 264
       240
        24
        24
         0
```

2.
```
        27  8/14
  14 ) 386
       280
       106
        98
         8
```
$27 \times 14 = 378$
$378 + 8 = 386$

3.
```
        43
  22 ) 946
       880
        66
        66
         0
```
$43 \times 22 = 946$

4.
```
        20  1/10
  10 ) 201
       200
         1
```
$20 \times 10 = 200$
$200 + 1 = 201$

5.
```
        10  3/25
  25 ) 253
       250
         3
```
$10 \times 25 = 250$
$250 + 3 = 253$

6.
```
        13
  38 ) 494
       380
       114
       114
         0
```
$13 \times 38 = 494$

7. $735 \div 21 = 35$ trees

8. $\$143 \div \$11 = 13$ books

Systematic Review 22D

1.
$$\begin{array}{r} 23 \\ 12\,\overline{)276} \\ \underline{240} \\ 36 \\ \underline{36} \\ 0 \end{array}$$

2. $23 \times 12 = 276$

3.
$$\begin{array}{r} 29\frac{1}{24} \\ 24\,\overline{)697} \\ \underline{480} \\ 217 \\ \underline{216} \\ 1 \end{array}$$

4. $29 \times 24 = 696$
$696 + 1 = 697$

5.
$$\begin{array}{r} 26 \\ 3\,\overline{)78} \\ \underline{60} \\ 18 \\ \underline{18} \\ 0 \end{array}$$

6. $26 \times 3 = 78$

7.
$$\begin{array}{r} 221\frac{3}{4} \\ 4\,\overline{)887} \\ \underline{800} \\ 87 \\ \underline{80} \\ 7 \\ \underline{4} \\ 3 \end{array}$$

8. $221 \times 4 = 884$
$884 + 3 = 887$

9. $10 + 32 = 42$
$42 \div 2 = 21$
$21 \times 12 = 252$ sq ft

10. $28 \times 8 = 224$
$224 \div 2 = 112$ sq in

11. $40 \times 40 = 1{,}600$ sq mi

12. done

13. done

14. $120 \div 12 = 10$

15. $169 \div 12 = 14\frac{1}{12}$ ft

16. $25 \times 5{,}280 = 132{,}000$ ft

Systematic Review 22E

1.
$$\begin{array}{r} 14\frac{13}{17} \\ 17\,\overline{)251} \\ \underline{170} \\ 81 \\ \underline{68} \\ 13 \end{array}$$

2. $14 \times 17 = 238$
$238 + 13 = 251$

3.
$$\begin{array}{r} 16 \\ 41\,\overline{)656} \\ \underline{410} \\ 246 \\ \underline{246} \\ 0 \end{array}$$

4. $16 \times 41 = 656$

5.
$$\begin{array}{r} 5\frac{5}{8} \\ 8\,\overline{)45} \\ \underline{40} \\ 5 \end{array}$$

6. $5 \times 8 = 40$
$40 + 5 = 45$

7.
$$\begin{array}{r} 59\frac{2}{9} \\ 9\,\overline{)533} \\ \underline{450} \\ 83 \\ \underline{81} \\ 2 \end{array}$$

8. $59 \times 9 = 531$
$531 + 2 = 533$

9. $9 + 29 = 38$
$38 \div 2 = 19$
$19 \times 11 = 209$ sq ft

10. $5 \times 16 = 80$
$80 \div 2 = 40$ sq in

11. $21 \times 32 = 672$ sq yd

12. $24 \div 12 = 2$ ft

13. $10 \times 12 = 120$ in

14. $180 \div 12 = 15$ ft
15. $576 \div 18 = 32$ miles per gallon
16. $6 \times 12 = 72$ in
17. $2 \times 2,000 = 4,000$ lb
18. $4,000 \times 16 = 64,000$ oz
    $64,000 - 32 = 63,968$ oz

Systematic Review 22F

1.
$$
\begin{array}{r}
22 \\
22\overline{)484} \\
440 \\
\hline
44 \\
44 \\
\hline
0
\end{array}
$$

2. $22 \times 22 = 484$

3.
$$
\begin{array}{r}
67\frac{12}{14} \\
14\overline{)950} \\
840 \\
\hline
110 \\
98 \\
\hline
12
\end{array}
$$

4. $67 \times 14 = 938$
   $938 + 12 = 950$

5.
$$
\begin{array}{r}
6\frac{3}{6} \\
6\overline{)39} \\
36 \\
\hline
3
\end{array}
$$

6. $6 \times 6 = 36$
   $36 + 3 = 39$

7.
$$
\begin{array}{r}
50\frac{4}{7} \\
7\overline{)354} \\
350 \\
\hline
4
\end{array}
$$

8. $50 \times 7 = 350$
   $350 + 4 = 354$

9. $30 + 50 = 80$
   $80 \div 2 = 40$
   $40 \times 40 = 1,600$ sq ft

10. $11 \times 34 = 374$
    $374 \div 2 = 187$ sq in

11. $25 \times 9 = 225$ sq ft

12. $72 \div 12 = 6$ ft
13. $16 \times 12 = 192$ in
14. $216 \div 12 = 18$ ft
15. $144 \div 12 = 12$ dozen
16. $36 \div 4 = \$9$
17. $16 \div 2 = 8$ times
18. $60 \times 10 = 600$ mi

Lesson Practice 23A

1. done

2. done

3.
$$
\begin{array}{r}
(200) \\
(5)\overline{)(1,000)}
\end{array}
$$

4.
$$
\begin{array}{r}
218 \\
5\overline{)1090} \\
1000 \\
\hline
90 \\
50 \\
\hline
40 \\
40 \\
\hline
0
\end{array}
$$
$218 \times 5 = 1090$

5.
$$
\begin{array}{r}
(1,000) \\
(7)\overline{)(9,000)}
\end{array}
$$

6.
$$
\begin{array}{r}
1,354\frac{1}{7} \\
7\overline{)9479} \\
7000 \\
\hline
2479 \\
2100 \\
\hline
379 \\
350 \\
\hline
29 \\
28 \\
\hline
1
\end{array}
$$
$1,354 \times 7 = 9,478$
$9,478 + 1 = 9,479$

7. $2,368 \div 2 = 1,184$ groups
8. $5,385 \div 5 = 1,077$ customers

Lesson Practice 23B

1.
$$
\begin{array}{r}
(600) \\
(9)\overline{)(6{,}000)}
\end{array}
$$

2.
$$
\begin{array}{r}
673\ \frac{6}{9} \\
9\overline{)6063} \\
\underline{5400} \\
663 \\
\underline{630} \\
33 \\
\underline{27} \\
6
\end{array}
$$
$673 \times 9 = 6{,}057$
$6{,}057 + 6 = 6{,}063$

3.
$$
\begin{array}{r}
(1{,}000) \\
(3)\overline{)(3{,}000)}
\end{array}
$$

4.
$$
\begin{array}{r}
861\ \frac{1}{3} \\
3\overline{)2584} \\
\underline{2400} \\
184 \\
\underline{180} \\
4 \\
\underline{3} \\
1
\end{array}
$$
$861 \times 3 = 2{,}583$
$2{,}583 + 1 = 2{,}584$

5.
$$
\begin{array}{r}
(2{,}000) \\
(2)\overline{)(5{,}000)}
\end{array}
$$

6.
$$
\begin{array}{r}
2{,}375\ \frac{1}{2} \\
2\overline{)4751} \\
\underline{4000} \\
751 \\
\underline{600} \\
151 \\
\underline{140} \\
11 \\
\underline{10} \\
1
\end{array}
$$
$2{,}375 \times 2 = 4{,}750$
$4{,}750 + 1 = 4751$

7. $1{,}125 \div 9 = 125$ mothers

8. $2{,}268 \div 5 = 453\frac{3}{5}$ miles

Lesson Practice 23C

1.
$$
\begin{array}{r}
(1{,}000) \\
(4)\overline{)(6{,}000)}
\end{array}
$$

2.
$$
\begin{array}{r}
1{,}505\ \frac{2}{4} \\
4\overline{)6022} \\
\underline{4000} \\
2022 \\
\underline{2000} \\
22 \\
\underline{20} \\
2
\end{array}
$$
$1{,}505 \times 4 = 6{,}020$
$6{,}020 + 2 = 6{,}022$

3.
$$
\begin{array}{r}
(800) \\
(7)\overline{)(6{,}000)}
\end{array}
$$

4.
$$
\begin{array}{r}
876\ \frac{2}{7} \\
7\overline{)6134} \\
\underline{5600} \\
534 \\
\underline{490} \\
44 \\
\underline{42} \\
2
\end{array}
$$
$876 \times 7 = 6{,}132$
$6{,}132 + 2 = 6{,}134$

5.
$$
\begin{array}{r}
(300) \\
(9)\overline{)(3{,}000)}
\end{array}
$$

6.
$$
\begin{array}{r}
279\ \frac{5}{9} \\
9\overline{)2516} \\
\underline{1800} \\
716 \\
\underline{630} \\
86 \\
\underline{81} \\
5
\end{array}
$$
$279 \times 9 = 2{,}511$
$2{,}511 + 5 = 2{,}516$

7. $9{,}936 \div 6 = 1{,}656$ ladybugs

8. $5{,}280 \div 4 = 1{,}320$

$1{,}320 \div 7 = 188\frac{4}{7}$ feet per day

Systematic Review 23D

1.
$$384\frac{2}{4}$$
$$4\overline{)1538}$$
$$\underline{1200}$$
$$338$$
$$\underline{320}$$
$$18$$
$$\underline{16}$$
$$2$$

2. $384 \times 4 = 1,536$
$1,536 + 2 = 1,538$

3.
$$1453\frac{1}{3}$$
$$3\overline{)4360}$$
$$\underline{3000}$$
$$1360$$
$$\underline{1200}$$
$$160$$
$$\underline{150}$$
$$10$$
$$\underline{9}$$
$$1$$

4. $1,453 \times 3 = 4,359$
$4,359 + 1 = 4,360$

5.
$$18$$
$$51\overline{)918}$$
$$\underline{510}$$
$$408$$
$$\underline{408}$$

6. $18 \times 51 = 918$

7.
$$11\frac{4}{38}$$
$$38\overline{)422}$$
$$\underline{380}$$
$$42$$
$$\underline{38}$$
$$4$$

8. $11 \times 38 = 418$
$418 + 4 = 422$

9. done

10.
$$3736$$
$$\underline{\times \ \ 31}$$
$$^13^736$$
$$^1 2 \ \ 1$$
$$\underline{9198}$$
$$115,816$$

11.
$$2945$$
$$\underline{\times \ \ 63}$$
$$^12 11$$
$$6725$$
$$1^5 2 3$$
$$\underline{2440}$$
$$185,535$$

12. $600 \div 3 = 200$ yd

13. $48 \div 2 = 24$ qt

14. $16 \div 4 = 4$ gal

15. no; $485 \div 4 = 121\frac{1}{4}$ cows

16. no

Systematic Review 23E

1.
$$250\frac{3}{8}$$
$$8\overline{)2003}$$
$$\underline{1600}$$
$$403$$
$$\underline{400}$$
$$3$$

2. $250 \times 8 = 2,000$
$2,000 + 3 = 2,003$

3.
$$692\frac{1}{2}$$
$$2\overline{)1385}$$
$$\underline{1200}$$
$$185$$
$$\underline{180}$$
$$5$$
$$\underline{4}$$
$$1$$

4. $692 \times 2 = 1,384$
$1,384 + 1 = 1,385$

5.
$$
\begin{array}{r}
39 \\
12\overline{)468} \\
360 \\
\hline 108 \\
108 \\
\hline 0
\end{array}
$$

6. $39 \times 12 = 468$

7.
$$
\begin{array}{r}
51\frac{6}{10} \\
10\overline{)516} \\
500 \\
\hline 16 \\
10 \\
\hline 6
\end{array}
$$

8. $51 \times 10 = 510$
$510 + 6 = 516$

9.
$$
\begin{array}{r}
1\ 3\ 4\ 8 \\
\times\ \ \ \ 29 \\
\hline {}^2 2^1 3\ 7 \\
9\ 7\ 6\ 2 \\
1 \\
{}^1 2\ 6\ 8\ 6 \\
\hline 3\ 9,0\ 9\ 2
\end{array}
$$

10.
$$
\begin{array}{r}
7\ 5\ 3\ 9 \\
\times\ \ \ \ 12 \\
\hline 1\ \ \ 1 \\
{}^1 4\ 4^1 0\ 6\ 8 \\
7\ 5\ 3\ 9 \\
\hline 9\ 0,4\ 6\ 8
\end{array}
$$

11.
$$
\begin{array}{r}
2\ 0\ 0\ 0 \\
\times\ \ \ \ 56 \\
\hline 1\ 2\ 0\ 0\ 0 \\
1\ 0\ 0\ 0\ 0 \\
\hline 1\ 1\ 2,0\ 0\ 0
\end{array}
$$

12. $444 \div 4 = \$111$
13. $720 \div 12 = 60$ ft
14. $80 \div 16 = 5$ lb
15. $2,067 \div 3 = 689$ bugs per hour
16. $4 \times 7 = 28$ qt
$28 \times 52 = 1,456$ qt per year

Systematic Review 23F

1.
$$
\begin{array}{r}
1,255\frac{1}{3} \\
3\overline{)3766} \\
3000 \\
\hline 766 \\
600 \\
\hline 166 \\
150 \\
\hline 16 \\
15 \\
\hline 1
\end{array}
$$

2. $1,255 \times 3 = 3,765$
$3,765 + 1 = 3,766$

3.
$$
\begin{array}{r}
1,498\frac{1}{6} \\
6\overline{)8989} \\
6000 \\
\hline 2989 \\
2400 \\
\hline 589 \\
540 \\
\hline 49 \\
48 \\
\hline 1
\end{array}
$$

4. $1,498 \times 6 = 8,988$
$8,988 + 1 = 8,989$

5.
$$
\begin{array}{r}
5\frac{45}{71} \\
71\overline{)400} \\
355 \\
\hline 45
\end{array}
$$

6. $5 \times 71 = 355$
$355 + 45 = 400$

7.
$$
\begin{array}{r}
8 \\
44\overline{)352} \\
352 \\
\hline 0
\end{array}
$$

8. $8 \times 44 = 352$

9.
$$
\begin{array}{r}
3\,4\,4\,5 \\
\times\ \ 9\,3 \\
\hline
^1\!1\ 1\ 1 \\
9\,2\,2\,5 \\
^1\!2^2\!3\ \ 3\,4 \\
7\,6\,6\,5 \\
\hline
3\,2\,0{,}3\,8\,5
\end{array}
$$

10.
$$
\begin{array}{r}
2\,3\,1\,7 \\
\times\ \ 6\,4 \\
\hline
^1\!1\ \ \ 2 \\
8\,2\,4\,8 \\
^1\!1\ \ \ 4 \\
1\,2\,8\,6\,2 \\
\hline
1\,4\,8{,}2\,8\,8
\end{array}
$$

11.
$$
\begin{array}{r}
8\,9\,1\,2 \\
\times\ \ 2\,5 \\
\hline
4\ \ \ 1 \\
^1\!4\ 0^1\!5\ 5\,0 \\
^1\!1\ 1 \\
6\,8\,2\,4 \\
\hline
2\,2\,2{,}8\,0\,0
\end{array}
$$

12. $5{,}280 \div 3 = 1{,}760$ yd

13. $4 \times 2{,}000 = 8{,}000$ lb

14. $1{,}000 \times 12 = 12{,}000$ in

15. $1{,}760 \times 2 = 3{,}520$ yd

16. perpendicular

Lesson Practice 24A

1. done

2. done

3.
$$
\begin{array}{r}
(200) \\
(20)\overline{)(4{,}000)}
\end{array}
$$

4.
$$
\begin{array}{r}
2\,0\,0 \\
1\,9\,\overline{)3\,8\,0\,0} \\
\underline{3\,8\,0\,0}
\end{array}
\qquad
\begin{array}{r}
1\,9 \\
\times\,2\,0\,0 \\
\hline
3\,8\,0\,0
\end{array}
$$

5.
$$
\begin{array}{r}
(100) \\
(60)\overline{)(7{,}000)}
\end{array}
$$

6.
$$
\begin{array}{r}
1\,2\,5\ \tfrac{45}{55} \\
5\,5\,\overline{)6\,9\,2\,0} \\
\underline{5\,5\,0\,0} \\
1\,4\,2\,0 \\
\underline{1\,1\,0\,0} \\
3\,2\,0 \\
\underline{2\,7\,5} \\
4\,5
\end{array}
\qquad
\begin{array}{r}
5\,5 \\
\times\,1\,2\,5 \\
\hline
2\,7\,5 \\
1\,1\,0 \\
5\,5 \\
\hline
6\,8\,7\,5 \\
+\ \ 4\,5 \\
\hline
6{,}9\,2\,0
\end{array}
$$

7. $\$6{,}929 \div 41 = 169$ hours

8. $\$1{,}875 \div \$15 = 125$ coats

Lesson Practice 24B

1. done

2. done

3.
$$
\begin{array}{r}
(40) \\
(90)\overline{)(4{,}000)}
\end{array}
$$

4.
$$
\begin{array}{r}
4\,2\ \tfrac{41}{91} \\
9\,1\,\overline{)3\,8\,6\,3} \\
\underline{3\,6\,4\,0} \\
2\,2\,3 \\
\underline{1\,8\,2} \\
4\,1
\end{array}
\qquad
\begin{array}{r}
9\,1 \\
\times\,4\,2 \\
\hline
1\,8\,2 \\
3\,6\,4 \\
\hline
3\,8\,2\,2 \\
+\ \ 4\,1 \\
\hline
3{,}8\,6\,3
\end{array}
$$

5.
$$
\begin{array}{r}
(200) \\
(30)\overline{)(6{,}000)}
\end{array}
$$

6.
$$
\begin{array}{r}
1\,6\,4\ \tfrac{16}{34} \\
3\,4\,\overline{)5\,5\,9\,2} \\
\underline{3\,4\,0\,0} \\
2\,1\,9\,2 \\
\underline{2\,0\,4\,0} \\
1\,5\,2 \\
\underline{1\,3\,6} \\
1\,6
\end{array}
\qquad
\begin{array}{r}
3\,4 \\
\times\,1\,6\,4 \\
\hline
1\,3\,6 \\
2\,0\,4 \\
3\,4 \\
\hline
5\,5\,7\,6 \\
+\ \ 1\,6 \\
\hline
5{,}5\,9\,2
\end{array}
$$

7. $4{,}275 \div 45 = 95$ boxes

8. $1{,}008 \div 14 = 72$ groups

Lesson Practice 24C

1.
$$\frac{(600)}{(10)\overline{|(6,000)}}$$

2.
$$509\frac{5}{12}$$
$$12\overline{|6113}$$
$$\underline{6000}$$
$$113$$
$$\underline{108}$$
$$5$$

$$12$$
$$\times 509$$
$$108$$
$$\underline{600}$$
$$6108$$
$$\underline{+\quad 5}$$
$$6,113$$

3.
$$\frac{(50)}{(80)\overline{|(4,000)}}$$

4.
$$55\frac{20}{75}$$
$$75\overline{|4145}$$
$$\underline{3750}$$
$$395$$
$$\underline{375}$$
$$20$$

$$75$$
$$\times 55$$
$$375$$
$$\underline{375}$$
$$4125$$
$$\underline{+\quad 20}$$
$$4,145$$

5.
$$\frac{(100)}{(50)\overline{|(7,000)}}$$

6.
$$152\frac{11}{46}$$
$$46\overline{|7003}$$
$$\underline{4600}$$
$$2403$$
$$\underline{2300}$$
$$103$$
$$\underline{92}$$
$$11$$

$$46$$
$$\times 152$$
$$92$$
$$230$$
$$\underline{46}$$
$$6992$$
$$\underline{+\quad 11}$$
$$7,003$$

7. $7,000 \div 35 = \$200$

8. $9,360 \div 52 = 180$ balls per week

Systematic Review 24D

1.
$$222\frac{3}{25}$$
$$25\overline{|5553}$$
$$\underline{5000}$$
$$553$$
$$\underline{500}$$
$$53$$
$$\underline{50}$$
$$3$$

2.
$$25$$
$$\times 222$$
$$1$$
$$140$$
$$140$$
$$40$$
$$5550$$
$$\underline{+\quad 3}$$
$$5,553$$

3.
$$95$$
$$63\overline{|5985}$$
$$\underline{5670}$$
$$315$$
$$\underline{315}$$
$$0$$

4.
$$63$$
$$\times\ 95$$
$$315$$
$$20$$
$$\underline{547}$$
$$5985$$

5.
$$363\frac{6}{8}$$
$$8\overline{|2910}$$
$$\underline{2400}$$
$$510$$
$$\underline{480}$$
$$30$$
$$\underline{24}$$
$$6$$

6.
```
        8
    × 3 6 3
      2 4
    4 8
    2 4
    2 9ᴵ0 4
    +   6
    2,9 1 0
```

7.
```
        9  9/11
  11 )108
      99
       9
```

8.
```
     1 1
   ×   9
      9 9
   +   9
    1 0 8
```

9. done

10.
```
          3 2 4
      ×  3 1 5 2
          6 4 8
        1 ᴵ1 2
      1 5 0 0
      3 2 4
      1
     ᴵ9 6 2
    1,0 2 1,2 4 8
```

11.
```
          7 9 3
      ×  1 2 8 4
          ᴵ3 1
        ²2 8 6 2
          7 2
       ²5 6 2 4
      ²1 1
        4 8 6
      7 9 3
    1,0 1 8,2 1 2
```

12. $12 \times 14 = 168$ sq ft

13. $16 + 14 + 7 + 20 + 8 = 65$
$65 \div 5 = 13$ oz < 1 lb

14. $7 \times 12 = 84$ in

Systematic Review 24E

1.
```
          87 33/34
   34 )2991
      2720
       271
       238
        33
```

2.
```
        3 4
      × 8 7
        2 2
      3 1 8
      2 4 2
     2 9ᴵ5 8
    +   3 3
    2,9 9 1
```

3.
```
          5 4 8  5/12
   12 )6 5 8 1
       6 0 0 0
         5 8 1
         4 8 0
         1 0 1
           9 6
            5
```

4.
```
        1 2
      × 5 4 8
      ᴵ 9 6
        4 8
      6 0
      6 5 7 6
    +     5
    6,5 8 1
```

5.
```
          9 2 6  3/5
    5 )4 6 3 3
       4 5 0 0
         1 3 3
         1 0 0
          3 3
          3 0
           3
```

6.
```
        5
    ×926
      30
      10
      45
    4630
    +  3
    4,633
```

7.
```
         18  9/27
    27 | 495
         270
         225
         216
           9
```

8.
```
        27
      ×18
      216
       27
      486
    +   9
      495
```

9.
```
         162
      × 4195
          31
         500
        ¹5 1
         948
       ²1 6 2
       2
       448
     679,590
```

10.
```
         281
      ×2703
           2
         643
       ⁴1 5
        4¹6 7
       1
       462
     759,543
```

11.
```
         437
      × 1588
        ¹2 5
       ¹3 2 46
        2 5
       ²3 2 46
        1 3
       2 0 5 5
       4 3 7
     6 9 3,9 5 6
```

12. $10 \times 16 = 160$

$160 \div 2 = 80$ sq in

13. $2,266 \div 11 = 206$ miles

14. E, F, H, M, N or Z

(possibly I, depending on style)

Systematic Review 24F

1.
```
          115  3/14
    14 | 1613
         1400
          213
          140
           73
           70
            3
```

2.
```
         14
      × 115
         70
         14
         14
       1610
     +    3
      1,613
```

3.
```
          13  40/91
    91 | 1223
         910
         313
         273
          40
```

4.
```
      91
    ×13
    273
   ¹91
   1183
   + 40
  1,223
```

5.
```
   1,890 2/4
  4)7562
   4000
   3562
   3200
    362
    360
      2
```

6.
```
       4
   ×1890
      36
      32
       4
    7560
   +   2
   7,562
```

7.
```
      22 8/13
  13)294
     260
      34
      26
       8
```

8.
```
     13
   ×22
     26
     26
   2¹86
   +  8
    294
```

9.
```
       561
   × 1361
     ¹561
     ¹¹3
    3066
      1
   ¹1583
    561
   763,521
```

10.
```
        946
     ×3722
        1
      ²¹882
        1
      ³1882
     ¹624
      382
    ¹211
     728
   3,521,012
```

11.
```
        837
     × 4693
        2
      ²²¹491
       26
     ²7273
      14
    ²4882
     12
    3228
   3,928,041
```

12. $5 + 7 = 12$
$12 \div 2 = 6$
$6 \times 12 = 72$ sq in

13. $20 \div 8 = 2$ r. 4; 2 full vans and 1
partial load = <u>3</u> vans, <u>no</u>

14. E, F, H, L, T; possibly I and J

Lesson Practice 25A

1. done

2. done

3.
```
       (1,000)
  (60))(60,000)
```

4.

$$1,054 \frac{14}{56}$$

$$56\overline{)59038}$$
$$56000$$
$$3038$$
$$2800$$
$$238$$
$$224$$
$$14$$

$$\begin{array}{r} 56 \\ \times 1054 \\ \hline 224 \\ {}^12\,30 \\ 50 \\ 56 \\ \hline 59024 \\ +\quad 14 \\ \hline 59,038 \end{array}$$

5.
$$(1,000)$$
$$(70)\overline{)(70,000)}$$

6.
$$73\overline{)71905}$$
$$65700$$
$$6205$$
$$5840$$
$$365$$
$$365$$
$$0$$

$$\begin{array}{r} 73 \\ \times 985 \\ \hline {}^13\,1 \\ {}^15\,2\,55 \\ {}^16\,2\,64 \\ 3\,7 \\ \hline 71,905 \end{array}$$

7. $\$40,000 \div \$20 = 2,000$

Lesson Practice 25B

1. done

2. done

3.
$$(1,000)$$
$$(500)\overline{)(800,000)}$$

4.
$$1,637 \frac{125}{503}$$
$$503\overline{)823536}$$
$$503000$$
$$320536$$
$$301800$$
$$18736$$
$$15090$$
$$3646$$
$$3521$$
$$125$$

$$\begin{array}{r} 503 \\ \times 1637 \\ \hline 2 \\ {}^13\,501 \\ {}^11\,509 \\ 3018 \\ 503 \\ \hline 823,411 \\ +\quad 125 \\ \hline 823,536 \end{array}$$

5. $963,600 \div 365 = 2,640$ feet per day

6. $\$26,052 \div 52 = \$501$ per week

Lesson Practice 25C

1.
$$(2,000)$$
$$(40)\overline{)(80,000)}$$

2.
$$2,160 \frac{25}{35}$$
$$35\overline{)75625}$$
$$70000$$
$$5625$$
$$3500$$
$$2125$$
$$2100$$
$$25$$

$$\begin{array}{r} 35 \\ \times 2160 \\ \hline {}^11\,30 \\ 8 \\ 35 \\ 10 \\ 6 \\ \hline 75600 \\ +\quad 25 \\ \hline 75,625 \end{array}$$

3.
$$(1,000)$$
$$(900)\overline{)(900,000)}$$

4.
$$1085 \frac{533}{858}$$
$$858\overline{)931463}$$
$$858000$$
$$73463$$
$$68640$$
$$4823$$
$$4290$$
$$533$$

$$\begin{array}{r} 858 \\ \times 1085 \\ \hline 4^12\,40 \\ {}^26\,465 \\ 404 \\ {}^18\,5\,8 \\ \hline 9\,30930 \\ +\quad 533 \\ \hline 9\,31,463 \end{array}$$

5. $12,560 \div 16 = 785$ lb

6. $900,000 \div 300 = 3,000$ days

Systematic Review 25D

1.
$$1,668 \frac{29}{38}$$
$$38\overline{)63413}$$
$$38000$$
$$25413$$
$$22800$$
$$2613$$
$$2280$$
$$333$$
$$304$$
$$29$$

2.
```
        38
    ×1,668
     ¹26
    ²1 444
    ²1 488
     388
       8
   63384
   +   29
   63,413
```

3.
```
         2,766  152
                357
    357 987614
        714000
        273614
        249900
         23714
         21420
          2294
          2142
           152
```

4.
```
       357
    ×2766
    ²134
      802
    ²134
      802
    234
     159
    11
    604
   987462
   +   152
   987,614
```

5.
```
      1,032  3
             9
    9 9291
      9000
       291
       270
        21
        18
         3
```

6.
```
         9
    ×1032
        18
        27
         9
      9288
    +    3
      9,291
```

7.
```
        569  3
             12
    12 6831
       6000
        831
        720
        111
        108
          3
```

8.
```
        12
    × 569
         1
    ¹198
    162
     50
    6828
    +   3
    6,831
```

9. $525 \div 15 = 35$ in

10. $28 \div 7 = 4$ ft

11. $500 \div 25 = 20$ in

12. $500 \times 12 = 6,000$ mi

13. $4,800 \div 12 = 400$ mph

14. $4,800 + 6,000 = 10,800$ mi

Systematic Review 25E

1.
$$1{,}094 \frac{60}{63}$$
$$63\overline{)68982}$$
$$\underline{63000}$$
$$5982$$
$$\underline{5670}$$
$$312$$
$$\underline{252}$$
$$60$$

2.
$$63$$
$$\times 1094$$
^{121}
$$5242$$
$$47$$
$$\underline{63}$$
$$68922$$
$$\underline{+\quad 60}$$
$$68{,}982$$

3.
$$2{,}157 \frac{39}{321}$$
$$321\overline{)692436}$$
$$\underline{642000}$$
$$50436$$
$$\underline{32100}$$
$$18336$$
$$\underline{16050}$$
$$2286$$
$$\underline{2247}$$
$$39$$

4.
$$321$$
$$\times 2157$$
$$21$$
^{11147}
$$505$$
$$321$$
$$\underline{642}$$
$$692397$$
$$\underline{+\quad 39}$$
$$692{,}436$$

5.
$$1{,}660$$
$$2\overline{)3320}$$
$$\underline{2000}$$
$$1320$$
$$\underline{1200}$$
$$120$$
$$\underline{120}$$
$$0$$

6.
$$2$$
$$\times 1660$$
$$12$$
$$12$$
$$\underline{2}$$
$$3{,}320$$

7.
$$183 \frac{39}{52}$$
$$52\overline{)9555}$$
$$\underline{5200}$$
$$4355$$
$$\underline{4160}$$
$$195$$
$$\underline{156}$$
$$39$$

8.
$$52$$
$$\times 183$$
^{1156}
$$416$$
$$0$$
$$\underline{52}$$
$$9516$$
$$\underline{+\quad 39}$$
$$9{,}555$$

9. $860 \div 20 = 43$ in

10. $88 \div 11 = 8$ ft

11. $255 \div 17 = 15$ in

12. $360 \times 2 = 720$ eggs
$720 \div 12 = 60$ dozen

13. $143 \div 11 = 13$ ft

14. $\$1{,}843 - \$792 = \$1{,}051$

Systematic Review 25F

1.

$$1,437\frac{21}{31}$$

$$31\overline{)44568}$$
$$\underline{31000}$$
$$13568$$
$$\underline{12400}$$
$$1168$$
$$\underline{930}$$
$$238$$
$$\underline{217}$$
$$21$$

2.

$$31$$
$$\underline{\times 1437}$$
$$217$$
$$93$$
$$1^12\,4$$
$$3\,1$$
$$\overline{44547}$$
$$\underline{+\quad 21}$$
$$44,568$$

3.

$$600$$
$$420\overline{)252000}$$
$$\underline{252000}$$
$$0$$

4.

$$420$$
$$\underline{\times 600}$$
$$12000$$
$$240000$$
$$252,000$$

5.

$$710\frac{3}{8}$$
$$8\overline{)5683}$$
$$\underline{5600}$$
$$83$$
$$\underline{80}$$
$$3$$

6.

$$8$$
$$\underline{\times 710}$$
$$8$$
$$56$$
$$\overline{5680}$$
$$\underline{+\quad 3}$$
$$5,683$$

7.

$$88\frac{12}{18}$$
$$18\overline{)1596}$$
$$\underline{1440}$$
$$156$$
$$\underline{144}$$
$$12$$

8.

$$18$$
$$\underline{\times 88}$$
$$6$$
$$^1{}^16\,8\,4$$
$$\underline{8\,4}$$
$$1\,5\,8\,4$$
$$\underline{+\quad 12}$$
$$1,5\,96$$

9. $110 \div 55 = 2$ in

10. $10,560 \div 132 = 80$ ft

11. $868 \div 31 = 28$ in

12. $560 \div 16 = 35$ lb

13. $2,000 \times 16 = 32,000$ oz

14. $543 + 324 + 480 = 1,347$

$1,347 \div 3 = 449$ mi per day

Lesson Practice 26A

1. done

2. $(5 \times 4) \times 2 = 40$ cubic units

3. $(3 \times 3) \times 2 = 18$ cubic inches

4. $(3 \times 4) \times 2 = 24$ cubic feet

5. $(6 \times 2) \times 3 = 36$ cu in

6. $(2 \times 8) \times 1 = 16$ cu ft

7. $(3 \times 3) \times 7 = 63$ cu in

8. $(3 \times 3) \times 3 = 27$ cu yd

9. $(10 \times 12) \times 4 = 480$ cu in

10. $(2 \times 3) \times 5 = 30$ cu ft

Lesson Practice 26B

1. $(2 \times 2) \times 2 = 8$ cu cm

2. $(5 \times 2) \times 2 = 20$ cu units

3. $(8 \times 3) \times 4 = 96$ cu in

4. $(4 \times 10) \times 2 = 80$ cu ft

5. $(5 \times 2) \times 3 = 30$ cu in

6. $(3 \times 7) \times 2 = 42$ cu ft

7. $(2 \times 2) \times 6 = 24$ cu in

8. $(4 \times 4) \times 4 = 64$ cu yd
9. $(4 \times 3) \times 3 = 36$ cu yd
10. $(5 \times 6) \times 3 = 90$ blocks

Lesson Practice 26C

1. $(3 \times 2) \times 1 = 6$ cu cm
2. $(4 \times 3) \times 3 = 36$ cu units
3. $(6 \times 3) \times 5 = 90$ cu in
4. $(4 \times 6) \times 3 = 72$ cu ft
5. $(6 \times 2) \times 4 = 48$ cu in
6. $(3 \times 8) \times 4 = 96$ cu ft
7. $(5 \times 5) \times 8 = 200$ cu in
8. $(9 \times 9) \times 9 = 729$ cu yd
9. $(4 \times 7) = 28$ gallons
10. $(5 \times 2) \times 1 = 10$ cu ft
$10 \times 7 = 70$ gal

Systematic Review 26D

1. $12 \times 12 \times 12 = 1,728$ cu ft
2. $4 \times 9 \times 5 = 180$ cu in

3.
$$1,306 \frac{2}{5}$$
$$5 \overline{)6532}$$
$$5000$$
$$1532$$
$$1500$$
$$32$$
$$30$$
$$2$$

4.
$$5$$
$$\times 1306$$
$$30$$
$$15$$
$$5$$
$$6530$$
$$+\quad 2$$
$$6,532$$

5.
$$2,460 \frac{4}{36}$$
$$36 \overline{)88564}$$
$$72000$$
$$16564$$
$$14400$$
$$2164$$
$$2160$$
$$4$$

6.
$$36$$
$$\times 2460$$
^{13}
$$1286$$
$$124$$
$$62$$
$$88560$$
$$+\quad 4$$
$$88,564$$

7. $10,560 \div 3 = 3,520$ yd
8. $100 \times 3 = 300$ ft
9. $300 \times 12 = 3,600$ in
10. $35 \div 7 = 5$ cu ft
11. $84 \times 7 = 588$ gal
12. $10 \times 4 \times 5 = 200$ cu in
$10 \times 6 \times 3 = 180$ cu in
$200 > 180$
13. $\$32 \div 4 = \$8$
14. $576 \div 24 = 24$ days
15. $10 \times 16 = 160$ oz

Systematic Review 26E

1. $10 \times 10 \times 10 = 1,000$ cu ft
2. $7 \times 11 \times 3 = 231$ cu in

3.
$$16$$
$$18 \overline{)288}$$
$$180$$
$$108$$
$$108$$
$$0$$

4.
$$18$$
$$\times 16$$
^{14}
$$168$$
$$8$$
$$288$$

5.
$$1{,}202\frac{245}{248}$$
$$248\overline{)298341}$$
$$\underline{248000}$$
$$50341$$
$$\underline{49600}$$
$$741$$
$$\underline{496}$$
$$245$$

6.
$$248$$
$$\times 1202$$
$$1$$
^{1486}
^{1486}
$$248$$
$$\overline{298^0^96}$$
$$+\quad 245$$
$$\overline{298{,}341}$$

7. $8 \times 2{,}000 = 16{,}000$ lb

8. $400 \div 16 = 25$ lb

9. $60 \times 16 = 960$ oz

10. $7 \times 6 \times 2 = 84$ cu ft
$84 \times 7 = 588$ gal

11. $324 \div 4 = 81$ fireflies

12. $15 \times 16 = 240$ oz
$240 > 179$; Joe gained more.

13. $\$549{,}600 \div 12 = \$45{,}800$ per month

14. $\$65{,}600 - \$45{,}800 = \$19{,}800$

15. $10 \times 8 = 80$; $80 \div 2 = 40$ sq in

Systematic Review 26F

1. $16 \times 5 \times 4 = 320$ cu ft

2. $10 \times 11 \times 3 = 330$ cu in

3.
$$2{,}710$$
$$6\overline{)16260}$$
$$\underline{12000}$$
$$4260$$
$$\underline{4200}$$
$$60$$
$$\underline{60}$$
$$0$$

4.
$$6$$
$$\times 2710$$
$$6$$
$$42$$
$$\underline{12}$$
$$16{,}260$$

5.
$$7{,}180\frac{11}{13}$$
$$13\overline{)93351}$$
$$\underline{91000}$$
$$2351$$
$$\underline{1300}$$
$$1051$$
$$\underline{1040}$$
$$11$$

6.
$$13$$
$$\times 7180$$
$$2$$
$$^1 84$$
$$2\quad 3$$
$$71$$
$$\overline{93340}$$
$$+\quad 11$$
$$\overline{93{,}351}$$

7. $18 \div 2 = 9$ qt

8. $75 \times 4 = 300$ quarters

9. $64 \times 4 = 256$ qt

10. $25{,}000 \div 5 = 5{,}000$ eggs

11. $10 \times 10 \times 5 = 500$ cu ft

12. $500 \times 7 = 3{,}500$ gal

13. $3{,}500 \times 8 = 28{,}000$ lb

14. yes

15. $14 + 8 = 22$
$22 \div 2 = 11$
$11 \times 10 = 110$ sq in

Lesson Practice 27A

1. done

2. 6 blocks
2 equal parts;
count 1 part
$\frac{1}{2}$ of 6 is 3

3. 10 blocks

5 equal parts;

count 4 parts

$\dfrac{4}{5}$ of 10 is 8.

4. 9 blocks

3 equal parts;

count 2 parts

$\dfrac{2}{3}$ of 9 is 6.

5. 8 blocks

4 equal parts;

count 3 parts

$\dfrac{3}{4}$ of 8 is 6.

Lesson Practice 27B

1. done

2. $\dfrac{1}{2}$ of 18 is 9

3. $\dfrac{5}{6}$ of 12 is 10

4. $20 \div 5 = 4$

$4 \times 3 = 12$

5. $6 \div 3 = 2$

$2 \times 2 = 4$

6. $8 \div 2 = 4$

$4 \times 1 = 4$

7. $6 \div 3 = 2$

$2 \times 1 = 2$

8. $8 \div 4 = 2$

$2 \times 1 = 2$

9. $10 \div 5 = 2$

$2 \times 4 = 8$

10. $6 \div 2 = 3$

$3 \times 1 = 3$

11. $12 \div 3 = 4$

$4 \times 2 = 8$

12. $8 \div 2 = 4$

$4 \times 1 = 4$

13. $12 \div 4 = 3$

$3 \times 1 = 3$ months

14. $10 \div 2 = 5$

$5 \times 1 = 5$ fingers

15. $10 \div 5 = 2$

$2 \times 2 = 4$ bulbs

Lesson Practice 27C

1. $\dfrac{1}{3}$ of 9 is 3

2. $\dfrac{3}{5}$ of 10 is 6

3. $\dfrac{2}{4}$ of 12 is 6

4. $16 \div 4 = 4$

$4 \times 2 = 8$

5. $10 \div 5 = 2$

$2 \times 1 = 2$

6. $18 \div 6 = 3$

$3 \times 5 = 15$

7. $12 \div 4 = 3$

$3 \times 3 = 9$

8. $10 \div 5 = 2$

$2 \times 3 = 6$

9. $4 \div 2 = 2$

$2 \times 1 = 2$

10. $12 \div 3 = 4$

$4 \times 1 = 4$

11. $12 \div 2 = 6$

$6 \times 1 = 6$

12. $16 \div 8 = 2$

$2 \times 7 = 14$

13. $14 \div 7 = 2$

$2 \times 2 = 4$ roses

14. $20 \div 5 = 4$

$4 \times 1 = 4$ bandages

15. $16 \div 4 = 4$

$4 \times 3 = 12$ years

Systematic Review 27D

1. $6 \div 2 = 3$

$3 \times 1 = 3$

2. $16 \div 4 = 4$

$4 \times 3 = 12$

3. $20 \div 5 = 4$

$4 \times 2 = 8$

4. $18 \times 6 \times 5 = 540$ cu ft

5. $12 \times 14 \times 2 = 336$ cu in

6. $6,003 \frac{3}{4}$

$4\overline{)24015}$
$\underline{24000}$
15
$\underline{12}$
3

7. $6,003 \times 4 = 24,012$
$24,012 + 3 = 24,015$

8. $2,322$

$21\overline{)48762}$
$\underline{42000}$
6762
$\underline{6300}$
462
$\underline{420}$
42
$\underline{42}$
0

9. $2,322 \times 21 = 48,762$
10. $5,280 \times 2 = 10,560$
11. $2,000 \times 4 = 8,000$
12. $75 \times 12 = 900$
13. $18 \div 3 = 6$
$6 \times 1 = 6$ girls
14. $3 \times 3 \times 4 = 36$ cu ft
$36 \times 7 = 252$ gal
15. $252 \times 8 = 2,016$ lb

Systematic Review 27E

1. $12 \div 6 = 2$
$2 \times 1 = 2$
2. $21 \div 7 = 3$
$3 \times 4 = 12$
3. $16 \div 8 = 2$
$2 \times 1 = 2$
4. $20 \times 11 \times 7 = 1,540$ cu ft
5. $6 \times 9 \times 1 = 54$ cu in

6. $3,278$

$9\overline{)29502}$
$\underline{27000}$
2502
$\underline{1800}$
702
$\underline{630}$
72
$\underline{72}$
0

7. $3,278 \times 9 = 29,502$

8. $2,031 \frac{3}{44}$

$44\overline{)89367}$
$\underline{88000}$
1367
$\underline{1320}$
47
$\underline{44}$
3

9. $2,031 \times 44 = 89,364$
$89,364 + 3 = 89,367$
10. $5 \times 3 = 15$
11. $7 \times 2 = 14$
12. $20 \times 4 = 80$
13. $9 \div 9 = 1$
$1 \times 3 = 3$ boys
14. $24 \div 6 = 4$
$4 \times 1 = 4$ hours per day
15. $4 \times 60 = 240$ minutes

Systematic Review 27F

1. $20 \div 4 = 5$
$5 \times 3 = 15$
2. $12 \div 3 = 4$
$4 \times 1 = 4$
3. $14 \div 2 = 7$
$7 \times 1 = 7$
4. $31 \times 15 \times 8 = 3,720$ cu ft
5. $10 \times 12 \times 3 = 360$ cu in

6.
$$
\begin{array}{r}
3,006 \\
6\,\overline{\smash{\big)}\,18036} \\
\underline{18000} \\
36 \\
\underline{36} \\
0
\end{array}
$$

7. $3,006 \times 6 = 18,036$

8.
$$
\begin{array}{r}
1,417 \\
25\,\overline{\smash{\big)}\,35425} \\
\underline{25000} \\
10425 \\
\underline{10000} \\
425 \\
\underline{250} \\
175 \\
\underline{175} \\
0
\end{array}
$$

9. $1,417 \times 25 = 35,425$
10. $20 \times 16 = 320$
11. $35 \times 4 = 140$
12. $5,280 \times 3 = 15,840$
13. $8,227,332 + 3,305,006 +$
    $4,884,234 = 16,416,572$ people
14. $8,274,961 - 7,477,503 = 797,458$ people
15. $8,274,961 + 7,477,503 = 15,752,464$
    $15,752,464 - 14,445,000 =$
    $1,307,464$ people

Lesson Practice 28A
1. done
2. 23
3. 17
4. 50
5. 200
6. 90
7. 70
8. 46
9. done
10. XIV
11. XXXI
12. XLVIII
13. CL
14. CCCXXV

15. CCXLIX
16. LXIII
17. 10:00
18. XIV
19. I,X,C
20. L,V

Lesson Practice 28B
1. 9
2. 36
3. 124
4. 152
5. 98
6. 192
7. 145
8. 315
9. XVIII
10. XXVI
11. XCIV
12. XLIII
13. CCLVIII
14. CCCXVII
15. CCLXII
16. CLXXXIX
17. 4:00
18. subtract
19. no
20. VI

Lesson Practice 28C
1. 19
2. 49
3. 94
4. 53
5. 147
6. 18
7. 39
8. 222
9. XCIV
10. XIII
11. XXVIII
12. XLV

13. CCCLIX
14. CVIII
15. CCCXI
16. CCLXXIV
17. 9:00
18. 20
19. 6:00
20. 3

Systematic Review 28D

1. 8
2. 24
3. 140
4. 42
5. LXI
6. XLVIII
7. CLII
8. CCX
9. $12 \div 3 = 4$
$4 \times 1 = 4$
10. $21 \div 7 = 3$
$3 \times 2 = 6$
11. $10 \div 5 = 2$
$2 \times 4 = 8$
12. $12 \times 45 = 540$
$540 \div 2 = 270$ sq in

13.
$$\begin{array}{r} 8{,}763 \tfrac{4}{7} \\ 7\,\overline{)61345} \\ \underline{56000} \\ 5345 \\ \underline{4900} \\ 445 \\ \underline{420} \\ 25 \\ \underline{21} \\ 4 \end{array}$$

14. $8{,}763 \times 7 = 61{,}341$

15.
$$\begin{array}{r} 2{,}014 \\ 425\,\overline{)855950} \\ \underline{850000} \\ 5950 \\ \underline{4250} \\ 1700 \\ \underline{1700} \\ 0 \end{array}$$

16. $2{,}014 \times 425 = 855{,}950$
17. $24 \div 3 = 8$
$8 \times 1 = 8$ children
18. $7 + 9 + 11 + 1 = 28$
$28 \div 4 = 7$ books

Systematic Review 28E

1. 24
2. 271
3. 145
4. 18
5. LXXV
6. XCII
7. CCCLXXX
8. CXI
9. $16 \div 4 = 4$
$4 \times 1 = 4$
10. $18 \div 6 = 3$
$3 \times 5 = 15$
11. $20 \div 2 = 10$
$10 \times 1 = 10$
12. $8 + 14 = 22$
$22 \div 2 = 11$
$11 \times 11 = 121$ sq ft

13.
$$\begin{array}{r} 1{,}528 \tfrac{1}{9} \\ 9\,\overline{)13753} \\ \underline{9000} \\ 4753 \\ \underline{4500} \\ 253 \\ \underline{180} \\ 73 \\ \underline{72} \\ 1 \end{array}$$

14. $1,528 \times 9 = 13,752$
$13,752 + 1 = 13,753$

15.
$$350 \overline{)784596} \quad 2,241 \frac{246}{350}$$
$$\underline{700000}$$
$$84596$$
$$\underline{70000}$$
$$14596$$
$$\underline{14000}$$
$$596$$
$$\underline{350}$$
$$246$$

16. $350 \times 2,241 = 784,350$
$784,350 + 246 = 784,596$

17. $12 + 11 + 2 + 3 = 28$
$28 \div 4 = 7$ in

18. $5 \times 12 = 60$
$60 + 7 = 67$ in

Systematic Review 28F

1. 140
2. 354
3. 27
4. 81
5. XXXIV
6. LVI
7. CCXCIX
8. CCCLV
9. $9 \div 3 = 3$
$3 \times 2 = 6$
10. $15 \div 5 = 3$
$3 \times 3 = 9$
11. $12 \div 4 = 3$
$3 \times 1 = 3$
12. $65 \times 65 = 4,225$ sq mi

13.
$$3 \overline{)33591} \quad 11,197$$
$$\underline{30000}$$
$$3591$$
$$\underline{3000}$$
$$591$$
$$\underline{300}$$
$$291$$
$$\underline{270}$$
$$21$$
$$\underline{21}$$
$$0$$

14. $11,197 \times 3 = 33,591$

15.
$$48 \overline{)639854} \quad 13,330 \frac{14}{48}$$
$$\underline{480000}$$
$$159854$$
$$\underline{144000}$$
$$15854$$
$$\underline{14400}$$
$$1454$$
$$\underline{1440}$$
$$14$$

16. $13,330 \times 48 = 639,840$
$639,840 + 14 = 639,854$

17. parallel

18. $12 \div 4 = 3$
$3 \times 3 = 9$ months

Lesson Practice 29A

1. done
2. $\frac{1}{3}$
3. $\frac{4}{5}$
4. $\frac{1}{2}$
5. $\frac{3}{4}$
6. $\frac{1}{6}$
7. $\frac{2}{5}$
8. $\frac{2}{3}$
9. denominator is 6; 6 pieces

Lesson Practice 29B

1. done

2.

3.

4.

5.

6.

7.

8.

9. numerator is 1; 1 piece

Lesson Practice 29C

1. done

2.

3.

4.

5.

6.

7.

8.

15.
```
        1,061
  38 ‾40318‾
      38000
       2318
       2280
         38
         38
          0
```

16. $1,061 \times 38 = 40,318$

17.   perpendicular

18. $15 \div 3 = 5$
$1 \times 5 = 5$ jelly beans

Systematic Review 29D

1. $\dfrac{3}{6}$

2. $\dfrac{1}{4}$

3.   9

4.   17

5.   250

6.   99

7.   XCI

8.   LIV

9.   LXIII

10.   CCCXCI

11. $12 \times 32 = 384$ sq in

12. $21 \times 40 = 840$
$840 \div 2 = 420$ sq in

13.
```
        116 6/17
  17 ‾1978‾
      1700
       278
       170
       108
       102
         6
```

14. $116 \times 17 = 1,972$
$1,972 + 6 = 1,978$

Systematic Review 29E

1.

2.

3.   72

4.   34

5.   194

6.   65

7.   XXV

8.   XIII

9.   XLVIII

10.   CLX

11. $16 \times 10 \times 5 = 800$ cu ft

12. $72 \times 91 \times 23 = 150,696$ cu in

13.
```
        53 14/65
  65 ‾3459‾
      3250
       209
       195
        14
```

14. $65 \times 53 = 3,445$
$3,445 + 14 = 3,459$

15.
$$21 \overline{)55627} \quad 2,648 \frac{19}{21}$$
$$\underline{42000}$$
$$13627$$
$$\underline{12600}$$
$$1027$$
$$\underline{840}$$
$$187$$
$$\underline{168}$$
$$19$$

16. $2,648 \times 21 = 55,608$
$55,608 + 19 = 55,627$

17. $200 \div 50 = 4$ hours

18. $12 \div 4 = 3$
$3 \times 3 = 9$ for Naomi
$12 - 9 = 3$ for Ruth

Systematic Review 29F

1.

2.

3. 14
4. 86
5. 209
6. 198
7. XXXVI
8. XIV
9. LIX
10. CCLXXIII
11. $100 \times 50 \times 30 = 150,000$ cu ft
12. $19 \times 24 \times 9 = 4,104$ cu in

13.
$$42 \overline{)7682} \quad 182 \frac{38}{42}$$
$$\underline{4200}$$
$$3482$$
$$\underline{3360}$$
$$122$$
$$\underline{84}$$
$$38$$

14. $182 \times 42 = 7,644$
$7,644 + 38 = 7,682$

15.
$$25 \overline{)60000} \quad 2,400$$
$$\underline{50000}$$
$$10000$$
$$\underline{10000}$$
$$0$$

16. $2,400 \times 25 = 60,000$

17. $350 \div 60 = 5 \frac{50}{60}$; 5 hours, 50 minutes

18. $50 \div 2 = 25$
$25 \times 1 = 25$ people

Lesson Practice 30A

1. done
2. done
3. 400
4. 1,250
5. 45,000
6. 952
7. 700
8. 2,003
9. done
10. done
11. DLXXVIII
12. $\overline{V}$
13. MMCXLVI
14. MDCCCLXXII
15. $\overline{X}$
16. MMLXV
17. 1945
18. MCMLV

19. MDCCLXXVI

20. $\overline{V}$

Lesson Practice 30B

1. 556
2. 1,421
3. 549
4. 5,200
5. 931
6. 514
7. 2,000,000
8. 1,320
9. DI
10. DCXXV
11. $\overline{C}$
12. MMM
13. $\overline{CM}$
14. CDXXXII
15. MCCLXIII
16. MCMLXVII
17. 1861
18. DLXIII
19. MCDXCII
20. 1,555

Lesson Practice 30C

1. 584
2. 606
3. 945
4. 20,000
5. 1,140
6. 829
7. 50,000
8. 900,000
9. DLXXXII
10. CMLXXIII
11. MLIII
12. MMMCC
13. CDXLIV
14. MDX
15. MMMCMXCV

16. $\overline{V}$
17. 1799
18. MCMLXVI
19. CDLIX
20. 1066

Systematic Review 30D

1. 2,900
2. 534
3. 1,600
4. 955
5. LVII
6. CIX
7. DLXXX
8. MCDXI
9. $\dfrac{4}{5}$
10. $\dfrac{2}{3}$
11.
$$89 \overline{)9012} \quad 101\frac{23}{89}$$
$$\underline{8900}$$
$$112$$
$$\underline{89}$$
$$23$$
12. $101 \times 89 = 8,989$

    $8,989 + 23 = 9,012$
13.
$$16 \overline{)32160} \quad 2,010$$
$$\underline{32000}$$
$$160$$
$$\underline{160}$$
$$0$$
14. $2,010 \times 16 = 32,160$
15. perpendicular
16. $8 \times 9 = 72$

    $72 \div 2 = 36$ sq ft
17. $85 + 92 + 73 + 98 = 348$

    $348 \div 4 = 87$
18. $\$345 = \$128 = \$217$

Systematic Review 30E

1. 928
2. 414
3. 1,000,000
4. 80
5. XXIX
6. CCXCIX
7. MMCMXCIX
8. $\bar{V}$
9. $\frac{2}{5}$
10. $\frac{1}{2}$
11.
$$208\frac{1}{38}$$
$$38\overline{)7905}$$
$$\underline{7600}$$
$$305$$
$$\underline{304}$$
$$1$$
12. $208 \times 38 = 7,904$
$7,904 + 1 = 7,905$
13.
$$2,296\frac{126}{200}$$
$$200\overline{)459326}$$
$$\underline{400000}$$
$$59326$$
$$\underline{40000}$$
$$19326$$
$$\underline{18000}$$
$$1326$$
$$\underline{1200}$$
$$126$$
14. $2,296 \times 200 = 459,200$
$459,200 + 126 = 459,326$
15. $17 \times 4 = 68$ qt
16. $5 \times 4 = 20$ quarters
17. $15 \times 21 = 315$ sq ft
18. $32 \div 16 = 2$ lb

Systematic Review 30F

1. 2,310
2. 590
3. 50,000
4. 171

5. XVII
6. CDL
7. MMMCCLXIX
8. DCLXXII
9.
$$1,333\frac{49}{51}$$
$$51\overline{)68032}$$
$$\underline{51000}$$
$$17032$$
$$\underline{15300}$$
$$1732$$
$$\underline{1530}$$
$$202$$
$$\underline{153}$$
$$49$$
$1,333 \times 51 = 67,983$
$67,983 + 49 = 68,032$
10.
$$3,143\frac{83}{182}$$
$$182\overline{)572109}$$
$$\underline{546000}$$
$$26109$$
$$\underline{18200}$$
$$7909$$
$$\underline{7280}$$
$$629$$
$$\underline{546}$$
$$83$$
$3,143 \times 182 = 572,026$
$572,026 + 83 = 572,109$
11. $4 \times 2,000 = 8,000$ lb
12. $6 + 8 = 14$
$14 \div 2 = 7$
$7 \times 5 = 35$ sq ft
13. $24 \div 3 = 8$
$8 \times 2 = 16$ swallows
14. $5,280 \times 2 = 10,560$ ft
15. 3,000
16. $5 \times 3 = 15$ ft
$15 \times 12 = 180$ in
17. $315 \div 45 = 7$ hours
18. $5 \times 6 \times 3 = 90$ cu ft

APPLICATION AND ENRICHMENT SOLUTIONS

Application and Enrichment 1G

wind-up mouse

1. Final answer is 120 (given).
2. Final answer is 486.
3. Final answer is 280.
4. Final answer is 576.

Application and Enrichment 2G

space shuttle

1. 10
2. 9
3. 2
4. 5
5. 2
6. 5
7. 1
8. 10
9. 3
10. 5
11. 2
12. 2
13. 4
14. 3
15. 2
16. 1
17. 4
18. 5
19. 8
20. 7

Application and Enrichment 3G

boat

1. divide
2. multiply
3. divide
4. divide
5. multiply

Application and Enrichment 4G

music box figure

1. $5A = 15$ or $15 \div 5 = A$
   $A = 3$
2. $3B = 30$ or $30 \div 3 = B$
   $B = 10$
   Students may use any letter they like for the unknown.
3. $2D = 12$ or $12 \div 2 = D$
   $D = 6$
4. $8H = 80$ or $80 \div 8 = H$
   $H = 10$

Application and Enrichment 5G

1. line segment   —
2. point   •
3. ray   →
4. line   ↔
5. point
6. line segment
7. ray
8. line

Application and Enrichment 6G

clock

These answers may be in any order:
$8 \times 3, 6 \times 4, 4 \times 6, 3 \times 8$

Application and Enrichment 7G

6 inches $\times$ 4 inches = 24 square inches

Application and Enrichment 8G

airplane

1. Put a black X on the four shapes that do not have four sides.
2. Five parallelograms; they all have two sets of parallel sides.

3. Three rectangles; they all have four right angles.
4. One square; all four sides are the same length.
   The unmarked figure is a trapezoid.

Application and Enrichment 9G

1. 90° + 90° + 90° + 90° = 360°
   Yes
2. 3 × 45° = 135° or 45° + 45° + 45° = 135°
   Smaller angles may be added to find the measure of larger angles.

1. There are two obtuse angles.
2. There are two right angles.
3. Use the definitions to check the angles. They may be turned in any direction.
4. 90° − 75° = 15°, so D = 15°

Application and Enrichment 10G

hot air balloon

1. always
2. more likely
3. always
4. never
5. less likely

Application and Enrichment 11G

1. Smith:
     1 + 2 + 7 + 10 = 20; 20 ÷ 4 = 5
   Jones:
     4 + 5 + 6 = 15; 15 ÷ 3 = 5
   Smith = Jones
2. Chloe:
     6 + 7 + 8 = 21; 21 ÷ 3 = 7
   Tucker:
     1 + 2 + 12 = 15; 15 ÷ 3 = 5
   Chloe > Tucker

3. Timothy:
     1 + 2 + 3 = 6 ÷ 3 = 2
   Peter:
     0 + 2 + 10 = 12; 12 ÷ 3 = 4
   Timothy < Peter

number of row	1	2	3	4	5	6	7	8
number of boxes in that row	1	2	3	4	5	6	7	8
total number of boxes	1	3	6	10	15	21	28	36

1. They are the same.
2. Sample answers: Add the number of boxes in each new row to the total number of boxes in the previous rows; look at the bottom row and add a number that is one more each time:
   1 + 2 = 3, 3 + 3 = 6, 6 + 4 = 10, 10 + 5 = 15, etc. (There may be other ways to describe the patterns in the chart.)

number of triangles	1	2	3	4	5	6	7
number of toothpicks	3	5	7	9	11	13	15

3. 13 toothpicks
4. 21 toothpicks. Sample answers: Each new triangle needs two more toothpicks; double the number of triangles and add one to find the number of toothpicks needed. (Experimenting with this is more important than finding the exact answer without help.)

Application and Enrichment 12G

carousel

1. subtract
2. divide
3. multiply
4. add
5. divide

Application and Enrichment 13G

tractor

1. 16 new blocks
2. 24 new squares
3. 32 new squares
4. After the first step, skip count by 8.

Application and Enrichment 14G

1. 8 quadrilaterals
2. 1 trapezoid (is also a quadrilateral)
3. 6 parallelograms (are also quadrilaterals)
4. 4 rectangles (are also parallelograms)
5. 2 squares (are also rectangles)
6. the triangle on the bottom right

1. 6 circles total
2. 10 circles total
3. 15 circles total
4. Sample answer: Add the number of the step to the total number of previous circles to get the new total. (There may be other ways to describe the pattern.)

Application and Enrichment 15G

sky - blue; grass - green; castle and sun - yellow; castle door - brown;

taller buildings - orange; shorter buildings - tan

Use the chart to match names with the correct number of sides.

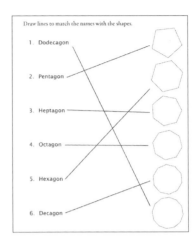

Application and Enrichment 16G

1. 2 squares
2. $18 \div 4 = 4$ r.2
3. Answers will vary.
4. 1 square
5. $16 \div 5 = 3$ r.1
6. Answers will vary.

1. $50 \div 7 = 7$ r.1
   Each neighbor gets 7 with 1 left over. It could be cut up and divided or given to someone else; Riley could also keep it. She could give it to one of the seven neighbors, but the shares would no longer be even.
2. $28 \div 5 = 5$ r.3
   Since pets cannot be cut up into pieces, he will need an extra cage for the remainder. Therefore, 6 cages are needed, but one cage will have 3 pets, not 5 pets.

3. 32 ft ÷ 3 ft = 10 pieces r. 2 (You must first change 1 yd to 3 ft.) Ten pieces are 3 ft or 1 yd long. The leftover piece is 2 ft long.

4. 17 ÷ 4 = 4 r.1
   Each sister gets 4 things if Julia wants to divide evenly. There is 1 thing left over. If it is something that could be cut, she could cut it into 4 equal pieces and give one to each sister.

5. 32 ÷ 6 = 5 r.2
   Jeff will need 6 shelves to hold all of the items in his collection. One shelf will have only 2 items.

Application and Enrichment 17G

Across

   1. quotient
   3. round
   4. base
   7. parallel
   10. rectangle
   11. right
   12. divisor
   13. perimeter

Down

   2. triangle
   5. factor
   6. areas
   8. average
   9. height

   1. 20, 20, 20, 20
   2. They are all the same.
   3. 1 + 19 = 20
   4. 6 + 14 = 20
   5. 2, 4, 6, 8 Skip count by 2.
   6. 10 – 0 = 10
   7. 40, 80, 120, 160
      Sample answers: Each number is 40 more than the one before;

skip count by 10 to find the second factors in the problems.
   8. 4 × 50 = 200
      4 × 60 = 240

Application and Enrichment 18G

Division problems
   5 ÷ 2 = 2 r.1
   17 ÷ 6 = 2 r.5
   39 ÷ 10 = 3 r.9
   19 ÷ 5 = 3 r.4
   55 ÷ 7 = 7 r.6
   39 ÷ 8 = 4 r.7
   20 ÷ 3 = 6 r.2
   53 ÷ 9 = 5 r.8
   35 ÷ 4 = 8 r.3
The numbers under the letters are (in order):
   1, 5, 9, 4, 6, 7, 2, 8, 3
Solution: I can do long division.

   1. small triangles: 8
      larger triangles with the sides of the square as bases: 4
      large triangle with diagonals as bases: 4
      8 + 4 + 4 = 16 triangles
   2. 10 squares
      16 triangles (see #1) + 16 triangles inside smaller square = 32 triangles

Application and Enrichment 19G

   1-4. done
   1. 60 miles
   2. 9 days
   3. 9 days
   4. no
   5. Day 7 on the graph; traveled for 2 days (7 – 5 = 2)
   6. no

Application and Enrichment 20G

1. top: 2, 4, 6, 8, 10, 12
   bottom: 7, 9, 11, 13, 15, 17
2. top: 1, 3, 9, 27, 81, 243
   bottom: 0, 2, 8, 26, 80, 242
3. top: 5, 10, 15, 20, 25, 30
   bottom: 6, 12, 18, 24, 30, 36
4. top: 2, 4, 6, 8, 10, 12
   (skip count by 2)
   bottom: 5, 7, 9, 11, 13, 15
   Sample answers: Add 3 to top number; add 2 to previous bottom number.
5. top: 3, 6, 9, 12, 15, 18, 21, 24
   (skip count by 3)
   bottom: 2, 5, 8, 11, 14, 17, 20, 23
   Sample answers: Subtract 1 from top number; add 3 to previous bottom number.

1. 2008
2. Country A
3. Production went down sharply and then began a steady increase.
4. no

Application and Enrichment 21G

This is called the "Haberdasher's Puzzle."

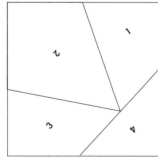

Application and Enrichment 22G

1. 6 ≥ 3; girls ate more
2. 8 − 4 = 4 more hot dogs
3. no

Ice Cream Cone Sales

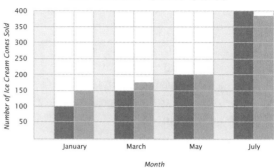

Application and Enrichment 23G

1.

2. 1 ft × 9 ft = 9 sq ft; 2 ft × 8 ft = 16 sq ft; 3 ft × 7 ft = 21 sq ft; 4 ft × 6 ft = 24 sq ft; 5 ft × 5 ft = 25 sq ft
3. The rectangle that is 1 unit × 9 units has the least area.
4. The rectangle that is 5 units × 5 units has the greatest area.

1. An extra pen will be needed.
2. Depending on what Sue made, she could divide the leftovers or save the remainder for herself or for some other purpose.
3. Write the remainder over the divisor to make a fraction.

Application and Enrichment 24G

no solutions

Application and Enrichment 25G

1. yes, yes, no, yes, yes
2. yes, for the purposes of this lesson (Most faces are slightly different on each side.)
3. no
4. Answers will vary.

1. top: 9, 18, 27, 36, 45, 54, 63, 72
   (skip count by 9)
   bottom: 3, 6, 9, 12, 15, 18, 21,
   24 (skip count by 3)
   Also, you can divide the top
   number by 3 to get the bottom
   number.
2. top: 2, 3, 4, 5, 6, 7, 8, 9
   (count by 1)
   middle: 4, 6, 8, 10, 12, 14, 16,
   18 (skip count by 2)
   bottom: 6, 9, 12, 15, 18, 21, 24,
   27 (skip count by 3)
   Also, each column skip counts by
   the number in the top row.
3. top: 20, 19, 18, 17, 16, 15, 14,
   13 (subtract 1 each time)
   bottom: 20, 21, 22, 23, 24, 25,
   26, 27 (add 1 each time)
   The difference between the top
   and bottom rows is 2 more each
   time.
4. Answers will vary.

Application and Enrichment 26G

The leaf is symmetrical around the line
of the fold.

Application and Enrichment 27G

Pan 1: (15" × 15") × 3" = 225" × 3" =
675 cubic inches
Pan 2: (12" × 12") × 3" = 144" × 3" =
432 cubic inches
Pan 3: (9" × 9") × 2" = 81" × 2" =
162 cubic inches
Pan 4: (6" × 6") × 4" = 36" × 4" =
144 cubic inches

675 cu in + 432 cu in + 162 cu in + 144
cu in = 1,413 cubic inches total volume

Answers will vary.

Application and Enrichment 28G

First line: 3, 6, 9, 12, 15, 18, 21, 24, 27
Second line: 1, 2, 3, 4, 5, 6, 7, 8, 9
Under letters: 1, 2, 3, 4, 5, 6, 7, 8, 9
Solution: All roads lead to Rome.

Unscrambled words in order:
square, triangle, trapezoid, area,
average, bases, height, ounces, divided,
sixteen

The triangle may be on any side of the
square and have any proportions, but
the overall shape must be a trapezoid.

Application and Enrichment 29G

horse and chariot

1. 55
2. 385
3. 160

Application and Enrichment 30G

1. 3 × 1,000 m = 3,000 m
2. 2 × 1,000 L = 2,000 mL
3. 500 ÷ 100 = 5 m

600 cm = 6 m
3,000 mL = 3 L
2,000 m = 2 km

TEST SOLUTIONS

Lesson Test 1
1. $4 \times 4 = 16$
2. $5 \times 3 = 15$
   $3 \times 5 = 15$
3. $3 \times \underline{9} = 27$
4. $8 \times \underline{8} = 64$
5. $4 \times \underline{5} = 20$
6. $6 \times \underline{1} = 6$
7. $5 \times \underline{9} = 45$
8. $6 \times \underline{4} = 24$
9. $10 \times \underline{10} = 100$
10. $2 \times \underline{7} = 14$
11. $5 \times 5 = 25$ sq ft
12. $8 \times 10 = 80$ sq in
13. $9 \times 7 = 63$ sq mi
14. $9 \times \underline{10} = \$90$; 10 hours
15. $8 \times 9 = 72$ sq ft

Lesson Test 2
1. $8 \div 1 = 8$
2. $18 \div 2 = 9$
3. $6 \div 2 = 3$
4. $4 \div 1 = 4$
5. $10 \div 2 = 5$
6. $8 \div 2 = 4$
7. $\frac{2}{2} = 1$
8. $\frac{7}{1} = 7$
9. $\frac{4}{2} = 2$
10. $5 \times \underline{3} = 15$
11. $5 \times \underline{5} = 25$
12. $2 \times \underline{10} = 20$
13. $10 \times \underline{6} = 60$
14. $8 \times 5 = 40$
15. $10 \times 7 = 70$
16. $4 \times 5 = 20$
17. $10 \times 3 = 30$
18. $10 \times 8 = 80$ sq in
19. $12 \div 2 = 6$ gumdrops
20. $8 \div 8 = 1$ piece each

Lesson Test 3
1. $20 \div 10 = 2$
2. $80 \div 10 = 8$
3. $60 \div 10 = 6$
4. $90 \div 10 = 9$
5. $8 \div 1 = 8$
6. $14 \div 2 = 7$
7. $16 \div 2 = 8$
8. $5 \div 1 = 5$
9. $6 \div 2 = 3$
10. $10 \div 2 = 5$
11. $10 \div 10 = 1$
12. $4 \div 2 = 2$
13. $5 \times 3 = 15$
14. $3 \times 3 = 9$
15. $6 \times 3 = 18$
16. $3 \times 4 = 12$
17. $10 \times 2 = 20$ pints
18. $8 \times 3 = 24$ ft
19. $\$40 \div \$10 = 4$ gifts
20. $7 \times 9 = 63$ sq in

Lesson Test 4
1. $15 \div 5 = 3$
2. $18 \div 3 = 6$
3. $12 \div 3 = 4$
4. $45 \div 5 = 9$
5. $30 \div 5 = 6$
6. $25 \div 5 = 5$
7. $27 \div 3 = 9$
8. $6 \div 3 = 2$
9. $14 \div 2 = 7$
10. $50 \div 10 = 5$
11. $24 \div 3 = 8$
12. $9 \div 3 = 3$
13. $3 \times \underline{6} = 18$
14. $7 \times \underline{3} = 21$
15. $5 \times \underline{4} = 20$

16.
$$\begin{array}{r} {}^1\!2\,5 \\ +\,3\,8 \\ \hline 6\,3 \end{array}$$

17.
$$\begin{array}{r} {}^1\!4\,7 \\ +\,7\,3 \\ \hline 1\,2\,0 \end{array}$$

18.
$$\begin{array}{r} {}^1\!6\,4 \\ +\,1\,9 \\ \hline 8\,3 \end{array}$$

19. $5 \times 7 = 35$ sq mi

20. $75 + 69 = 144$ beads

Lesson Test 5

1. $9 \times 9 = 81$
2. $9 \times 2 = 18$
3. $9 \times 5 = 45$
4. $9 \times 8 = 72$
5. $9 \times 6 = 54$
6. $9 \times 3 = 27$
7. $9 \times 7 = 63$
8. $9 \times 4 = 36$
9. $35 \div 5 = 7$
10. $24 \div 3 = 8$
11. $5 \div 5 = 1$
12. $18 \div 3 = 6$
13. $21 \div 3 = 7$
14. $25 \div 5 = 5$
15. $15 \div 3 = 5$
16. $16 \div 2 = 8$
17. $\parallel$
18. $\perp$
19. $30 \div 3 = 10$ yd
20. $\$15 + \$26 = \$41$

Lesson Test 6

1. $18 \div 9 = 2$
2. $36 \div 9 = 4$
3. $45 \div 9 = 5$
4. $81 \div 9 = 9$
5. $27 \div 9 = 3$
6. $54 \div 9 = 6$
7. $72 \div 9 = 8$

8. $63 \div 9 = 7$
9. $90 \div 10 = 9$
10. $10 \div 2 = 5$
11. $15 \div 3 = 5$
12. $12 \div 3 = 4$
13. $24 \div 3 = 8$
14. $20 \div 2 = 10$
15. $9 \div 9 = 1$
16. $21 \div 3 = 7$

17.
$$\begin{array}{r} {}^8\!\!\not{9}\,\,{}^1\!1 \\ -\,7\,6 \\ \hline 1\,5 \end{array}$$

18.
$$\begin{array}{r} {}^3\!\!\not{4}\,\,{}^1\!2 \\ -\,1\,3 \\ \hline 2\,9 \end{array}$$

19.
$$\begin{array}{r} {}^7\!\!\not{8}\,\,{}^1\!0 \\ -\,3\,5 \\ \hline 4\,5 \end{array}$$

20. $18 + 17 = 35$
    $35 \div 5 = 7$ treats each

Unit Test I

1. $4 \div 2 = 2$
2. $24 \div 3 = 8$
3. $45 \div 9 = 5$
4. $2 \div 1 = 2$
5. $18 \div 3 = 6$
6. $30 \div 5 = 6$
7. $81 \div 9 = 9$
8. $60 \div 10 = 6$
9. $16 \div 2 = 8$
10. $21 \div 3 = 7$
11. $63 \div 9 = 7$
12. $5 \div 5 = 1$
13. $27 \div 9 = 3$
14. $10 \div 2 = 5$
15. $40 \div 10 = 4$
16. $12 \div 3 = 4$
17. $45 \div 9 = 5$
18. $14 \div 2 = 7$
19. $25 \div 5 = 5$
20. $18 \div 2 = 9$

21. $50 \div 5 = 10$
22. $9 \div 3 = 3$
23. $9 \div 9 = 1$
24. $27 \div 3 = 9$
25. $90 \div 10 = 9$
26. $15 \div 3 = 5$
27. $12 \div 2 = 6$
28. $4 \div 1 = 4$
29. $40 \div 5 = 8$
30. $54 \div 9 = 6$
31. $70 \div 10 = 7$
32. $15 \div 5 = 3$
33. $45 \div 5 = 9$
34. $30 \div 3 = 10$
35. $5 \div 1 = 5$
36. $100 \div 10 = 10$
37. $9 \div 1 = 9$
38. $20 \div 5 = 4$
39. $72 \div 9 = 8$
40. $20 \div 2 = 10$
41. $6 \div 3 = 2$
42. $10 \div 5 = 2$
43. $18 \div 9 = 2$
44. $35 \div 5 = 7$
45. $8 \div 2 = 4$
46. $80 \div 10 = 8$
47. $8 \div 1 = 8$
48. $3 \div 3 = 1$

49.
$$\begin{array}{r} {}^{1}5\,6 \\ +\,3\,9 \\ \hline 9\,5 \end{array}$$

50.
$$\begin{array}{r} {}^{5}\cancel{6}\,{}^{1}2 \\ -\,2\,5 \\ \hline 3\,7 \end{array}$$

51.
$$\begin{array}{r} {}^{7}\cancel{8}\,{}^{1}1 \\ -\,4\,6 \\ \hline 3\,5 \end{array}$$

52. $\parallel$
53. $\perp$
54. $8 \times 9 = 72$ sq ft
55. $5 \times 3 = 15$ ft
56. $27 \div 3 = 9$ yd
57. $6 \times 2 = 12$ pt
58. $24 \div 2 = 12$ qt

Lesson Test 7

1. $9 \times 10 = 90$ sq ft
2. $3 \times 7 = 21$ sq in
3. $36 \div 9 = 4$
4. $18 \div 3 = 6$
5. $20 \div 5 = 4$
6. $12 \div 2 = 6$
7. $81 \div 9 = 9$
8. $40 \div 10 = 4$
9. $4 \div 2 = 2$
10. $63 \div 9 = 7$
11. $6 \times \underline{2} = 12$
12. $6 \times \underline{5} = 30$
13. $6 \times \underline{9} = 54$
14. $6 \times \underline{3} = 18$
15. $6 \times \underline{7} = 42$
16. $6 \times \underline{8} = 48$
17. $6 \times \underline{4} = 24$
18. $6 \times \underline{6} = 36$
19. $25 - 16 = 9$
    $9 \div 3 = 3$ cages
20. $18 \div 2 = 9$ qt

Lesson Test 8

1. $12 \div 6 = 2$
2. $24 \div 6 = 4$
3. $54 \div 6 = 9$
4. $30 \div 6 = 5$
5. $42 \div 6 = 7$
6. $48 \div 6 = 8$
7. $18 \div 6 = 3$
8. $36 \div 6 = 6$
9. $72 \div 9 = 8$
10. $20 \div 5 = 4$
11. $8 \div 2 = 4$
12. $27 \div 3 = 9$

13.
$$\begin{array}{r} {}^{1}\cancel{2}\,{}^{1}3 \\ -\quad 5 \\ \hline 1\,8 \end{array}$$

14.
$$\begin{array}{r} {}^1\!72 \\ +19 \\ \hline 91 \end{array}$$

15.
$$\begin{array}{r} {}^4\!\cancel{5}\,{}^1\!3 \\ -\ 45 \\ \hline 8 \end{array}$$

16.
$$\begin{array}{r} 22 \\ \times13 \\ \hline 66 \\ 22 \\ \hline 286 \end{array}$$

17.
$$\begin{array}{r} 45 \\ \times24 \\ \hline {}^{1\,2} \\ 160 \\ 80 \\ \hline 1{,}080 \end{array}$$

18.
$$\begin{array}{r} 16 \\ \times37 \\ \hline {}^{1\,4} \\ 172 \\ 38 \\ \hline 592 \end{array}$$

19. $20 \div 2 = 10$ people

20. $36 \div 6 = 6$ ft

Lesson Test 9

1. $4 \times 5 = 20$
$20 \div 2 = 10$ sq in

2. $7 \times 6 = 42$ sq ft

3. $3 \times 2 = 6$
$6 \div 2 = 3$ sq yd

4. $40 \div 10 = 4$

5. $12 \div 3 = 4$

6. $8 \div 2 = 4$

7. $45 \div 5 = 9$

8. $4 \times \underline{6} = 24$

9. $4 \times \underline{8} = 32$

10. $4 \times \underline{4} = 16$

11. $4 \times \underline{7} = 28$

12.
$$\begin{array}{r} 84 \\ \times22 \\ \hline {}^1\!68 \\ 168 \\ \hline 1{,}848 \end{array}$$

13.
$$\begin{array}{r} 43 \\ \times35 \\ \hline {}^1\!215 \\ 129 \\ \hline 1{,}505 \end{array}$$

14.
$$\begin{array}{r} 67 \\ \times54 \\ \hline {}^{1\ 2} \\ 248 \\ 335 \\ \hline 3{,}618 \end{array}$$

15. $25+15+24+61 = 125$

16. $44+38+62+56+11 = 211$

17. $90+23+57+18+82 = 270$

18. no

19. $25+16+18+32 = 91$ animals

20. $91 \times 4 = 364$ hooves

Lesson Test 10

1. $8 \div 4 = 2$

2. $32 \div 4 = 8$

3. $16 \div 4 = 4$

4. $12 \div 4 = 3$

5. $28 \div 4 = 7$

6. $20 \div 4 = 5$

7. $36 \div 4 = 9$

8. $24 \div 4 = 6$

9. $18 \div 6 = 3$

10. $42 \div 6 = 7$

11. $30 \div 6 = 5$

12. $48 \div 6 = 8$

13. $71+34+59+26 = 190$

14.
$$\begin{array}{r} {}^5\!\cancel{6}\,{}^1\!5 \\ -\ 39 \\ \hline 26 \end{array}$$

15.
$$
\begin{array}{r}
84 \\
\times 62 \\
\hline
168 \\
2 \\
^14\,84 \\
\hline
5{,}208
\end{array}
$$

16. $5 \times 4 = 20$
$20 \div 2 = 10$ sq ft
17. $32 \times 25 = 800$ sq in
18. $2 \times 5 = 10$
$10 \div 2 = 5$ sq yd
19. $24 \div 4 = 6$ gal
20. $40 \div 4 = \$10$

Lesson Test 11
1. $2+7+8+3 = 20$
$20 \div 4 = 5$
2. $4+3+6+7+10 = 30$
$30 \div 5 = 6$
3. $5+11+14 = 30$
$30 \div 3 = 10$
4. $7 \times 7 = 49$
5. $7 \times 6 = 42$
6. $7 \times 8 = 56$
7. $8 \times 8 = 64$
8. $8 \times 7 = 56$
9. $8 \times 9 = 72$
10. $63 \div 9 = 7$
11. $48 \div 6 = 8$
12. $40 \div 5 = 8$
13. $21 \div 3 = 7$
14. $32 \div 4 = 8$
15. $80 \div 10 = 8$
16. $7 \times 7 = 49$ sq ft
17. $1 \times 8 = 8$
$8 \div 2 = 4$ sq mi
18. yes
19. $11+16 = 27$ ft
$27 \div 3 = 9$ yd
20. $3+2+4+3 = 12$
$12 \div 4 = 3$ pt average
$12 \div 2 = 6$ qt total

Lesson Test 12
1. $28 \div 7 = 4$
2. $63 \div 7 = 9$
3. $56 \div 8 = 7$
4. $16 \div 8 = 2$
5. $14 \div 7 = 2$
6. $35 \div 7 = 5$
7. $24 \div 8 = 3$
8. $72 \div 8 = 9$
9. $48 \div 8 = 6$
10. $42 \div 7 = 6$
11. $21 \div 7 = 3$
12. $40 \div 8 = 5$
13. $49 \div 7 = 7$
14. $32 \div 8 = 4$
15. $64 \div 8 = 8$
16. $56 \div 7 = 8$
17. $7 \times 9 = 63$ sq in
18. $3 \times 16 = 48$ oz
$48 < 50$
50-oz bag is more.
19. $28 \div 4 = \$7$
20. $1+2+7+10+12+16 = 48$
$48 \div 6 = 8$

Unit Test II
1. $36 \div 4 = 9$
2. $24 \div 6 = 4$
3. $32 \div 8 = 4$
4. $49 \div 7 = 7$
5. $56 \div 8 = 7$
6. $20 \div 4 = 5$
7. $42 \div 6 = 7$
8. $60 \div 6 = 10$
9. $16 \div 8 = 2$
10. $35 \div 7 = 5$
11. $48 \div 8 = 6$
12. $7 \div 7 = 1$
13. $12 \div 4 = 3$
14. $36 \div 6 = 6$
15. $70 \div 7 = 10$
16. $12 \div 6 = 2$
17. $8 \div 8 = 1$

18. $56 \div 7 = 8$
19. $28 \div 4 = 7$
20. $42 \div 7 = 6$
21. $24 \div 8 = 3$
22. $32 \div 4 = 8$
23. $80 \div 8 = 10$
24. $21 \div 7 = 3$
25. $18 \div 6 = 3$
26. $4 \div 4 = 1$
27. $14 \div 7 = 2$
28. $72 \div 8 = 9$
29. $40 \div 4 = 10$
30. $64 \div 8 = 8$
31. $6 \div 6 = 1$
32. $48 \div 6 = 8$
33. $40 \div 8 = 5$
34. $24 \div 4 = 6$
35. $63 \div 7 = 9$
36. $30 \div 6 = 5$
37. $28 \div 7 = 4$
38. $8 \div 4 = 2$
39. $54 \div 6 = 9$
40. $16 \div 4 = 4$
41. $35 + 72 + 15 + 48 = 170$

42.
$$\begin{array}{r} {}^{8}\cancel{9}\;{}^{1}\cancel{1} \\ -\;3\;6 \\ \hline 5\;5 \end{array}$$

43.
$$\begin{array}{r} 7\,5 \\ \times 5\,8 \\ \hline 4 \\ {}^{1}5\,6\,0 \\ 2 \\ {}^{1}3\,5\,5 \\ \hline 4,3\,5\,0 \end{array}$$

44. $7 \times 2 = 14$
$14 \div 2 = 7$ sq yd
45. $5 + 9 + 13 = 27$
$27 \div 3 = 9$
46. $6 \times 4 = 24$ qt
47. $32 \div 4 = 8$ gal
48. $9 \times 4 = 36$ quarters
49. $20 \div 4 = \$5$
50. $2 \times 16 = 32$ oz

Lesson Test 13

1. $8 + 12 = 20$
$20 \div 2 = 10$
$10 \times 6 = 60$ sq in
2. $2 \times 7 = 14$
$14 \div 2 = 7$ sq ft
3. $3 + 9 = 12$
$12 \div 2 = 6$
$6 \times 8 = 48$ sq ft
4. $52 \times 36 = 1,872$ sq in
5. $14 \times 20 = 280$
6. $22 \times 30 = 660$
7. $43 \times 30 = 1,290$
8. $51 \times 20 = 1,020$
9. $42 \div 7 = 6$
10. $100 \div 10 = 10$
11. $64 \div 8 = 8$
12. $27 \div 9 = 3$

13.
$$\begin{array}{r} 3\,5\,2 \\ +\,1\,2\,6 \\ \hline 4\,7\,8 \end{array}$$

14.
$$\begin{array}{r} {}^{7}\cancel{8}\;{}^{6}\cancel{9}\;{}^{1}1 \\ -\;3\;\;4\,9 \\ \hline 4\;\;6\,2 \end{array}$$

15.
$$\begin{array}{r} 1 \\ 6\,0\,7 \\ +\,7\,8\,5 \\ \hline 1,3\,9\,2 \end{array}$$

16. $8 \div 4 = 2$ gal
17. $28 \div 4 = \$7$
18. $27 \div 3 = 9$ yd
19. $50 \div 5 = 10$ ft
20. $\$315 - \$227 = \$88$

Lesson Test 14

1. $221,346$
2. $3,467,000$
3. $6,000,000 + 100,000 +$
$20,000 + 3,000 + 500$
4. $4,000,000 + 500,000$

5. $6 + 10 = 16$
   $16 \div 2 = 8$
   $8 \times 7 = 56$ sq in

6. $19 \times 25 = 475$ sq ft

7. $6 \times 200 = 1{,}200$

8. $24 \times 200 = 4{,}800$

9. $17 \times 100 = 1{,}700$

10. $32 \times 200 = 6{,}400$

11. $56 \div 8 = 7$

12. $49 \div 7 = 7$

13. $9 \div 9 = 1$

14. $24 \div 6 = 4$

15. 2

16. $5 \times 16 = 80$ oz

17. $8 \times 2 = 16$ sq in
    $16 \div 2 = 8$ sq in

18. $16 \div 2 = 8$ qt
    $8 \div 4 = 2$ gal

Lesson Test 15

1. 7,632,400,000; seven billion,
   six hundred thirty-two million,
   four hundred thousand

2. 555,431,000;
   five hundred fifty-five million,
   four hundred thirty-one thousand

3. 1,635,721,000,000

4. 4,315,021

5. $8 \times 1{,}000{,}000{,}000 +$
   $2 \times 100{,}000{,}000 + 5 \times 10{,}000{,}000$

6. $3 \times 1{,}000{,}000{,}000{,}000 +$
   $4 \times 100{,}000{,}000{,}000 +$
   $2 \times 100 + 7 \times 10 + 4 \times 1$

7. $10 \times 60 = 600$

8. $12 \times 40 = 480$

9. $20 \times 200 = 4{,}000$

10.
$$
\begin{array}{r}
{\scriptstyle 1\ 1} \\
2{,}5\,4\,3 \\
+\,8{,}0\,6\,7 \\
\hline
10{,}6\,1\,0
\end{array}
$$

11.
$$
\begin{array}{r}
6{,}4\,\overset{\scriptstyle 3}{\cancel{}}\,\overset{\scriptstyle 15}{\cancel{6}}\,{}^{1}0 \\
-\quad 1\ 9\ 2 \\
\hline
6{,}2\ \ 6\ 8
\end{array}
$$

12.
$$
\begin{array}{r}
{\scriptstyle 1\ \ 2\ 1} \\
1{,}2\,4\,7 \\
3{,}5\,9\,8 \\
+\,6{,}0\,1\,3 \\
\hline
10{,}8\,5\,8
\end{array}
$$

13. $63 \div 9 = 7$

14. $36 \div 6 = 6$

15. $21 \div 7 = 3$

16. $8 \div 8 = 1$

17. $3 + 5 + 11 + 13 = 32$
    $32 \div 4 = 8$

18. $16 \times 12 = 192$ hours

Lesson Test 16

1.
$$
\begin{array}{r}
3\,\text{r.1} \\
3\,\overline{)10} \\
\underline{9} \\
1
\end{array}
$$

2.
$$
\begin{array}{r}
4\,\text{r.1} \\
6\,\overline{)25} \\
\underline{24} \\
1
\end{array}
$$

3.
$$
\begin{array}{r}
3\,\text{r.3} \\
9\,\overline{)30} \\
\underline{27} \\
3
\end{array}
$$

4.
$$
\begin{array}{r}
7\,\text{r.3} \\
4\,\overline{)31} \\
\underline{28} \\
3
\end{array}
$$

5.
$$
\begin{array}{r}
7\,\text{r.1} \\
2\,\overline{)15} \\
\underline{14} \\
1
\end{array}
$$

6.
$$
\begin{array}{r}
9\,\text{r.2} \\
5\,\overline{)47} \\
\underline{45} \\
2
\end{array}
$$

7.
$$\begin{array}{r} 2\ r.4 \\ 8\overline{)20} \\ \underline{16} \\ 4 \end{array}$$

8.
$$\begin{array}{r} 5\ r.6 \\ 7\overline{)41} \\ \underline{35} \\ 6 \end{array}$$

9.
$$\begin{array}{r} 4\ r.5 \\ 8\overline{)37} \\ \underline{32} \\ 5 \end{array}$$

10.
$$\begin{array}{r} {}^2\cancel{3},\!\cancel{0}^6\cancel{7}\,{}^1\!6 \\ -1,\!4\ 6\ 7 \\ \hline 1,\!6\ 0\ 9 \end{array}$$

11.
$$\begin{array}{r} 4,\!{}^5\cancel{6}\,{}^1\!4\,\cancel{5}\,{}^1\!4 \\ -3,\!2\ 9\ 8 \\ \hline 1,\!3\ 5\ 6 \end{array}$$

12.
$$\begin{array}{r} 6,\!512 \\ +7,\!285 \\ \hline 13,\!797 \end{array}$$

13. $11 \times 700 = 7,700$
14. $12 \times 300 = 3,600$
15. $21 \times 400 = 8,400$
16. 8,310,675,420
17. $18 \div 4 = 4\ r.\ 2$
     4 full cages with 2 left over
18. $45 \times 38 = 1,710$ sq ft

Lesson Test 17

1. 1,008; $800 + 200 + 8$
2. 1,008; $800 + 200 + 8$
3. $90 \div 3 = 30$
4. $240 \div 6 = 40$
5. $120 \div 2 = 60$
6. $200 \div 5 = 40$
7. $55 \div 6 = 9\ r.\ 1$
8. $26 \div 8 = 3\ r.\ 2$
9. $64 \div 9 = 7\ r.\ 1$
10. $30 \div 7 = 4\ r.\ 2$
11. $3 \times 16 = 48$ oz

12. $4 \times 4 = 16$ qt
13. $33 \div 3 = 11$ yd
14. $1 \times 2,000 = 2,000$ lb
15. $6 \times 2,000 = 12,000$ lb
16. $3 \times 2,000 = 6,000$ lb
17. $31 \div 4 = 7\ r.\ 3$
     $7 per child with $3 left over
18. 5,400,600,000,005

Lesson Test 18

1.
$$\begin{array}{r} 21 \\ 2\overline{)42} \\ \underline{40} \\ 2 \\ \underline{2} \\ 0 \end{array} \qquad \begin{array}{r} 2 \\ \times 21 \\ \hline 42 \end{array}$$

2.
$$\begin{array}{r} 10\ r.2 \\ 9\overline{)92} \\ \underline{90} \\ 2 \end{array} \qquad \begin{array}{r} 9 \\ \times 10 \\ \hline 90 \\ +\ 2 \\ \hline 92 \end{array}$$

3.
$$\begin{array}{r} 22\ r.1 \\ 3\overline{)67} \\ \underline{60} \\ 7 \\ \underline{6} \\ 1 \end{array} \qquad \begin{array}{r} 3 \\ \times 22 \\ \hline 66 \\ +\ 1 \\ \hline 67 \end{array}$$

4.
$$\begin{array}{r} 4\ r.3 \\ 4\overline{)19} \\ \underline{16} \\ 3 \end{array} \qquad \begin{array}{r} 4 \\ \times\ 4 \\ \hline 16 \\ +\ 3 \\ \hline 19 \end{array}$$

5.
$$\begin{array}{r} 4\ r.1 \\ 5\overline{)21} \\ \underline{20} \\ 1 \end{array} \qquad \begin{array}{r} 5 \\ \times\ 4 \\ \hline 20 \\ +\ 1 \\ \hline 21 \end{array}$$

6.
$$\begin{array}{r} 60 \\ 8\overline{)480} \\ \underline{480} \\ 0 \end{array} \qquad \begin{array}{r} 8 \\ \times 60 \\ \hline 480 \end{array}$$

7. ^{1}25
    175
    +45
    145

8. 8
    7
    4
    3
    +2
    24

9. ^{1} ^{1}95
    +345
    440

10. $4 \times 2{,}000 = 8{,}000$
11. $31 \times 4 = 124$
12. $10 \times 16 = 160$
13. $1 \times 5{,}280 = 5{,}280$
14. $3 \times 5{,}280 = 15{,}840$
15. $5 \times 5{,}280 = 26{,}400$
16. $4 + 8 + 15 + 25 = 52$
    $52 \div 4 = 13$
17. $25 \div 4 = 6$ r. 1
    6 pages filled, 1 picture left over
18. $5 \times 3 = 15$ ft

Lesson Test 19

1.   238
    3⟌714
        600
        114
         90
         24
         24
          0

2.   3
      ×238
        24
         9
         6
       714

3.   125 r.3
    5⟌628
        500
        128
        100
         28
         25
          3

4.   5
       ×125
         25
         10
          5
        625
      +   3
        628

5.   92
    4⟌368
        360
          8
          8
          0

6.   4
      ×92
       368

7.   16 r.1
    7⟌113
         70
         43
         42
          1

8.   7
      ×16
        42
         7
       112
      + 1
       113

9.   453
      × 46
        231
        408
    ^{1}1212
         60
      20,838

10.
```
      839
    ×  25
    4ᴴ1 4
    0 5 5
      1
   ᴴ1 6 6 8
   2 0,9 7 5
```

11.
```
      8 5 1
    ×   6 9
      ᴴ4
    7 2 5 9
      3
   ᴴ4 8 0 6
   5 8,7 1 9
```

12. $40 \times 3 = 120$

13. $2 \times 5{,}280 = 10{,}560$

14. $5 \times 16 = 80$

15. $2{,}495{,}000{,}000$

16. parallel

Lesson Test 20

1.
```
      179
   5 |895
     500
     395
     350
      45
      45
       0
       5
```

2.
```
   × 179
     895
```

3.
```
       44 4/8
   8 |356
     320
      36
      32
       4
```

4.
```
       8
     ×44
      32
      32
     352
    +  4
     356
```

5.
```
      204 2/3
   3 |614
     600
      14
      12
       2
```

6.
```
       3
     ×204
      12
       6
     612
    +  2
     614
```

7.
```
       96 2/6
   6 |578
     540
      38
      36
       2
```

8.
```
       6
     ×96
      36
     540
     576
    +  2
     578
```

9. $300 \div 5 = 60$ mi

10. $3 \times 5{,}280 = 15{,}840$ ft

11. $35 \div 7 = 5$ yards

12. base $= 7 \times 3 = 21$ ft
    height $= 5 \times 3 = 15$ ft
    $21 \times 15 = 315$ sq ft

13. $8 \times 2{,}000 = 16{,}000$ lb
    $16{,}000 > 1{,}600$
    8 tons is more weight.

14. 4

15. $2{,}003 - 1{,}974 = 29$ years

Lesson Test 21

1. 30
2. 100
3. 1,000
4. 10
5. 400
6. 5,000
7.
$$6\,\overline{)(900)}^{(100)}$$
8.
$$\begin{array}{r} 144\frac{5}{6} \\ 6\,\overline{)869} \\ 600 \\ \hline 269 \\ 240 \\ \hline 29 \\ 24 \\ \hline 5 \end{array}$$
9.
$$\begin{array}{r} 326 \\ 2\,\overline{)652} \\ 600 \\ \hline 52 \\ 40 \\ \hline 12 \\ 12 \\ \hline 0 \end{array}$$
10.
$$\begin{array}{r} 2 \\ \times 326 \\ \hline 12 \\ 4 \\ 6 \\ \hline 652 \end{array}$$
11.
$$\begin{array}{r} 34\frac{1}{7} \\ 7\,\overline{)239} \\ 210 \\ \hline 29 \\ 28 \\ \hline 1 \end{array}$$
12.
$$\begin{array}{r} 7 \\ \times 34 \\ \hline 28 \\ 21 \\ \hline 238 \\ + 1 \\ \hline 239 \end{array}$$
13. $2 \times 5,280 = 10,560$ ft
14. $5 \times 2,000 = 10,000$ lb
15. $5 \times 500 = 2,500$ mi
16. $13 + 17 = 30$
$30 \div 2 = 15$
$15 \times 11 = 165$ sq ft

Unit Test III

1. $5 + 19 = 24$
$24 \div 2 = 12$
$12 \times 6 = 72$ sq in
2. $2 + 18 = 20$
$20 \div 2 = 10$
$10 \times 5 = 50$ sq ft
3. 3,761,800,000
4. 2,413,283,350,000
5. $7 \times 10,000,000 + 5 \times 1,000,000 + 1 \times 100,000 + 2 \times 10,000 + 3 \times 1,000$
6. $28 \times 60 = 1,680$
7. $13 \times 400 = 5,200$
8. $56 \times 700 = 39,200$
9.
$$\begin{array}{r} 30 \\ 2\,\overline{)60} \\ 60 \\ \hline 0 \end{array}$$
10.
$$\begin{array}{r} 8\text{ r.}1 \\ 8\,\overline{)65} \\ 64 \\ \hline 1 \end{array}$$
11.
$$\begin{array}{r} 9\text{ r.}3 \\ 4\,\overline{)39} \\ 36 \\ \hline 3 \end{array}$$

12.
```
      223
  3 ) 669
      600
       69
       60
        9
        9
        0
```

13.
```
      51 r.2
  5 ) 257
      250
        7
        5
        2
```

14.
```
      103 r.5
  8 ) 829
      800
       29
       24
        5
```

15. 30

16. 200

17. 3,000

18.
```
      (100)
  5 ) (900)
```

19.
```
      181 3/5
  5 ) 908
      500
      408
      400
        8
        5
        3
```

20.
```
      216 1/4
  4 ) 865
      800
       65
       40
       25
       24
        1
```

21.
```
         4
    × 216
        24
         4
         8
       864
      +  1
       865
```

22.
```
       78 1/2
   2 ) 157
       140
        17
        16
         1
```

23.
```
         2
     × 78
        16
        14
       156
      +  1
       157
```

24. $5 \times 5{,}280 = 26{,}400$ ft

25. $9 \times 2{,}000 = 18{,}000$ lb

26. $568 \div 2 = 284$ people

Lesson Test 22

1.
```
        15 7/34
   34 ) 517
        340
        177
        170
          7
```

2. $15 \times 34 = 510$
$510 + 7 = 517$

3.
```
        20 7/18
   18 ) 367
        360
          7
```

4. $20 \times 18 = 360$
$360 + 7 = 367$

5.
$$15\frac{3}{5}$$
$$5\overline{)78}$$
$$\underline{50}$$
$$28$$
$$\underline{25}$$
$$3$$

6. $15 \times 5 = 75$
$75 + 3 = 78$

7.
$$103\frac{7}{9}$$
$$9\overline{)934}$$
$$\underline{900}$$
$$34$$
$$\underline{27}$$
$$7$$

8. $103 \times 9 = 927$
$927 + 7 = 934$

9. $20 + 40 = 60$
$60 \div 2 = 30$
$30 \times 30 = 900$ sq ft

10. $12 \times 59 = 708$
$708 \div 2 = 354$ sq in

11. 455 sq ft

12. $144 \div 12 = 12$

13. $20 \times 12 = 240$

14. $72 \div 12 = 6$

15. $48 \div 4 = \$12$

16. $330 \div 55 = 6$ hours

17. $330 \div 11 = 30$ miles

18. $7 \times 2,000 = 14,000$ lb
$140,000 > 14,000$

Lesson Test 23

1.
$$912\frac{2}{5}$$
$$5\overline{)4562}$$
$$\underline{4500}$$
$$62$$
$$\underline{50}$$
$$12$$
$$\underline{10}$$
$$2$$

2.
$$5$$
$$\times 912$$
$$10$$
$$5$$
$$\underline{4\ 5}$$
$$4,560 + 2 = 4,562$$

3.
$$213\frac{2}{7}$$
$$7\overline{)1493}$$
$$\underline{1400}$$
$$93$$
$$\underline{70}$$
$$23$$
$$\underline{21}$$
$$2$$

4.
$$7$$
$$\times 213$$
$$21$$
$$7$$
$$\underline{14}$$
$$1491$$
$$\underline{+\ \ 2}$$
$$1,493$$

5.
$$4\frac{30}{82}$$
$$82\overline{)358}$$
$$\underline{328}$$
$$30$$

6.
$$4$$
$$\times 82$$
$$8$$
$$\underline{32}$$
$$328$$
$$\underline{+30}$$
$$358$$

7.
$$40$$
$$21\overline{)840}$$
$$\underline{840}$$
$$0$$

8.
$$21$$
$$\times 40$$
$$840$$

9. 4,137
    × 13
    2
    12¹391
    4137
    53,781

10. 2,428
    × 75
    ¹¹2¹4
    0000
    215
    14846
    182,100

11. 7,801
    × 36
    4
    ¹42806
    2
    21403
    280,836

12. 396 ÷ 12 = 33 ft
13. 3 parallel lines
14. 80 ÷ 16 = 5 lb
    5 < 10; yes
15. 2,272 × 4 = 9,088 qt

Lesson Test 24

1. $108\frac{24}{25}$
    25)2724
    2500
    224
    200
    24

2. 25
    ×108
    ¹4
    160
    25
    2700
    + 24
    2,724

3. $28\frac{66}{81}$
    81)2334
    1620
    714
    648
    66

4. 81
    × 28
    1648
    162
    11
    2268
    + 66
    2,334

5. $1,445\frac{3}{6}$
    6)8673
    6000
    2673
    2400
    273
    240
    33
    30
    3

6. 6
    ×1445
    30
    24
    24
    6
    8670
    + 3
    8,673

7. 31
    12)372
    360
    12
    12
    0

8. 12
    ×31
    12
    36
    372

9.
$$\begin{array}{r} 3\,45 \\ \times\,1\,4\,7\,2 \\ \hline {}^16\ 1 \\ {}^1{}_2{}^12\ 3\ 8\ 0 \\ 1\ 8\ 5 \\ {}^11\ 1\ 2 \\ 2\ 6\ 0 \\ 3\ 4\ 5 \\ \hline 5\ 0\ 7,8\ 4\ 0 \end{array}$$

10.
$$\begin{array}{r} 8\,3\,7 \\ \times\,4\,7\,5\,4 \\ \hline {}^23\ 1\ 2 \\ 2\ 2\ 8 \\ {}^14\ 1\ 3 \\ 0\ 5\ 5 \\ {}^15\ 2\ 4 \\ 6\ 1\ 9 \\ 3\ 1\ 2 \\ 2\ 2\ 8 \\ \hline 3,9\ 7\ 9,0\ 9\ 8 \end{array}$$

11.
$$\begin{array}{r} 6\,3\,2 \\ \times\,3\,5\,8\,1 \\ \hline 6\ 3\ 2 \\ {}^24\ {}^12\ 1 \\ 8\ 4\ 6 \\ {}^13\ 1\ 1 \\ 0\ 5\ 0 \\ {}^11\ 8\ 9\ 6 \\ \hline 2,\ 2\ 6\ 3,1\ 9\ 2 \end{array}$$

12. $13 + 19 = 32$

$32 \div 2 = 16$

$16 \times 41 = 656$ sq ft

13. $34 \div 5 = 6$ r. 4

6 full cars and 1 partial car; 7 cars

14. ||

Lesson Test 25

1.
$$\begin{array}{r} 1{,}153\,\frac{19}{29} \\ 29\,\overline{)33456} \\ \underline{29000} \\ 4456 \\ \underline{2900} \\ 1556 \\ \underline{1450} \\ 106 \\ \underline{87} \\ 19 \end{array}$$

2.
$$\begin{array}{r} 29 \\ \times\,1\,1\,5\,3 \\ \hline 2 \\ {}_1{}_1\,6\ 7 \\ 1\ 4\ 5 \\ {}_1\,2\ 9 \\ 2\ 9 \\ \hline 3\ 3\ 4\ 3\ 7 \\ +\quad 1\ 9 \\ \hline 3\ 3,4\ 5\ 6 \end{array}$$

3.
$$\begin{array}{r} 2{,}081\,\frac{95}{180} \\ 180\,\overline{)374675} \\ \underline{360000} \\ 14675 \\ \underline{14400} \\ 275 \\ \underline{180} \\ 95 \end{array}$$

4.
$$\begin{array}{r} 1\,8\,0 \\ \times\,\ 2\,0\,8\,1 \\ \hline 6\ 1\ 8\ 0 \\ {}^16\ 8\ 4 \\ 2 \\ \hline 3\ 7\ 4\ {}^15\ 8\ 0 \\ +\quad\ 9\ 5 \\ \hline 3\ 7\ 4,6\ 7\ 5 \end{array}$$

5.
$$656$$
$$7\overline{)4592}$$
$$\underline{4200}$$
$$392$$
$$\underline{350}$$
$$42$$
$$\underline{42}$$
$$0$$

6.
$$7$$
$$\times 656$$
$$42$$
$$35$$
$$\underline{42}$$
$$4,592$$

7.
$$108\frac{11}{15}$$
$$15\overline{)1631}$$
$$\underline{1500}$$
$$131$$
$$\underline{120}$$
$$11$$

8.
$$15$$
$$\times 108$$
$$4$$
$$1^15\,80$$
$$1620$$
$$\underline{+\quad 11}$$
$$1631$$

9. $336 \div 12 = 28$ in
10. $37,500 \div 250 = 150$ ft
11. $323 \div 19 = 17$ in
12. $56 \times 16 = 896$ oz
13. $270 \div 6 = 45$ mph
14. $780 + 4,670 + 550 = 6,000$
$6,000 \div 3 = 2,000$ lb
$2,000$ lb $= 1$ ton

Lesson Test 26
1. $24 \times 12 \times 6 = 1,728$ cu ft
2. $33 \times 45 \times 10 = 14,850$ cu in

3.
$$5,474$$
$$5\overline{)27370}$$
$$\underline{25000}$$
$$2370$$
$$\underline{2000}$$
$$370$$
$$\underline{350}$$
$$20$$
$$\underline{20}$$
$$0$$

4.
$$5$$
$$\times 5\,474$$
$$20$$
$$35$$
$$20$$
$$\underline{25}$$
$$27,370$$

5.
$$2,570\frac{21}{32}$$
$$32\overline{)82261}$$
$$\underline{64000}$$
$$18261$$
$$\underline{16000}$$
$$2261$$
$$\underline{2240}$$
$$21$$

6.
$$32$$
$$\times 2570$$
$$21$$
$$^11\,1\,14$$
$$50$$
$$\underline{64}$$
$$82\,240$$
$$\underline{+\quad 21}$$
$$82,261$$

7. $28 \div 2 = 14$
8. $44 \times 4 = 176$
9. $48 \div 16 = 3$
10. $6 \times 2,000 = 12,000$
11. $100 \div 4 = 25$
12. $60 \div 12 = 5$
13. $20 \times 15 \times 6 = 1,800$ cu ft
14. $1,800 \times 7 = 12,600$ gal
15. $12,600 \times 8 = 100,800$ lb

Lesson Test 27

1. $21 \div 7 = 3$
   $3 \times 4 = 12$
2. $9 \div 3 = 3$
   $3 \times 2 = 6$
3. $36 \div 9 = 4$
   $4 \times 1 = 4$
4. $75 \div 5 = 15$
   $15 \times 3 = 45$
5. $18 \div 6 = 3$
   $3 \times 5 = 15$
6. $28 \div 2 = 14$
   $14 \times 1 = 14$
7.
```
        9,715 2/3
   3 |29147
       27000
        2147
        2100
          47
          30
          17
          15
           2
```
8.
```
          3
       ×9715
         15
          3
         21
         27
      29145
   +      2
      29,147
```
9.
```
        1,011 30/46
   46 |46536
       46000
         536
         460
          76
          46
          30
```

10.
```
          46
       × 1011
       ¹ 46
         46
      46
     46506
   +    30
     46,536
```
11. $36 \div 3 = 12$
12. $24 \div 4 = 6$
13. $8 \times 5,280 = 42,240$
14. $32 \div 4 = 8$
    $8 \times 1 = 8$ flowers
15. $100 \times 100 \times 90 = 900,000$ cu in

Lesson Test 28

1. 14
2. 72
3. 230
4. 99
5. XLI
6. LXXXV
7. CCCXXXIII
8. XXIX
9. $12 \div 2 = 6$
   $6 \times 1 = 6$
10. $64 \div 8 = 8$
    $8 \times 3 = 24$
11. $15 \div 5 = 3$
    $3 \times 2 = 6$
12. $11 + 25 = 36$
    $36 \div 2 = 18$
    $18 \times 13 = 234$ sq ft
13.
```
        5,373
   5 |26865
       25000
        1865
        1500
         365
         350
          15
          15
           0
```

14.
$$
\begin{array}{r}
5 \\
\times\,5\,373 \\
\hline
15 \\
35 \\
15 \\
25 \\
\hline
26,865
\end{array}
$$

15.
$$
2{,}284\,\dfrac{140}{216}
$$
$$
\begin{array}{r}
216\,\overline{)493484} \\
432000 \\
\hline
61484 \\
43200 \\
\hline
18284 \\
17280 \\
\hline
1004 \\
864 \\
\hline
140
\end{array}
$$

16.
$$
\begin{array}{r}
216 \\
\times\,2\,284 \\
\hline
2 \\
{}^{1}8\,44 \\
{}^{1\ 2}4 \\
1688 \\
1 \\
422 \\
1 \\
422 \\
\hline
493344 \\
+\quad 140 \\
\hline
493{,}484
\end{array}
$$

17. $16 + 43 + 58 + 91 = 208$
$208 \div 4 = 52$

18. $24 \div 6 = 4$
$4 \times 1 = 4$ horses

Lesson Test 29

1. $\dfrac{5}{6}$

2. $\dfrac{1}{3}$

3.

4.

5. 26

6. 43

7. 165

8. 192

9. XLVII

10. XVIII

11. CCXIX

12. CLIV

13.
$$
283\,\dfrac{20}{31}
$$
$$
\begin{array}{r}
31\,\overline{)8793} \\
6200 \\
\hline
2593 \\
2480 \\
\hline
113 \\
93 \\
\hline
20
\end{array}
$$

14.
$$
\begin{array}{r}
31 \\
\times\,283 \\
\hline
{}^{1}93 \\
248 \\
62 \\
\hline
8773 \\
+\quad 20 \\
\hline
8,793
\end{array}
$$

15.
$$
1{,}417\,\dfrac{7}{14}
$$
$$
\begin{array}{r}
14\,\overline{)19845} \\
14000 \\
\hline
5845 \\
5600 \\
\hline
245 \\
140 \\
\hline
105 \\
98 \\
\hline
7
\end{array}
$$

16.
```
        1 4
   × 1 4 1 7
      ₁ 2 8
      ₁ 1 4
      4 6
      1 4
   1 9 8¹3 8
   +      7
   1 9,8 4 5
```

17. $20 \times 9 = 180$
$180 \div 2 = 90$ sq in

18. $165 \div 55 = 3$ hr

Lesson Test 30

1. 2,200
2. 525
3. 750
4. 929
5. LVIII
6. DXX
7. MMMDCC
8. MCMLXV
9. $\frac{1}{5}$
10. $\frac{3}{3}$

11.
```
       127 ⁵/₆₄
   64⌐8133
      6400
      1733
      1280
       453
       448
         5
```

12.
```
        6 4
   × 1 2 7
   ¹4 4 8
   ¹1 2 8
   6 4
   8 1¹2 8
   +     5
   8.1 3 3
```

13.
```
        1,286
   500⌐643000
       500000
       143000
       100000
        43000
        40000
         3000
         3000
            0
```

14.
```
        5 0 0
   × 1 2 8 6
     3 0 0 0
   4 0 0 0 0
   1 0 0 0
   5 0 0
   6 4 3,0 0 0
```

15. $7 + 9 = 16$
$16 \div 2 = 8$
$8 \times 8 = 64$ sq ft

16. 600

17. $135 \div 15 = 9$ hours

18. $25 \div 5 = 5$
$5 \times 1 = 5$ lb spoiled
$25 - 5 = 20$ lb left

Unit Test IV

1.
```
       43 ⁹/₁₃
   13⌐568
      520
       48
       39
        9
```

2.
```
        1 3
   × 4 3
   1 3 9
   4 2
   5¹5 9
   +   9
   5 6 8
```

3.
$$32\frac{11}{30}$$
$$30\overline{)971}$$
$$\underline{900}$$
$$71$$
$$\underline{60}$$
$$11$$

4.
```
    30
  × 32
    60
   90
   960
 + 11
   971
```

5.
$$9{,}030\frac{4}{5}$$
$$5\overline{)45154}$$
$$\underline{45000}$$
$$154$$
$$\underline{150}$$
$$4$$

6.
```
      5
 × 9030
     15
   45
  45150
 +    4
  45154
```

7.
$$2{,}872\frac{4}{24}$$
$$24\overline{)68932}$$
$$\underline{48000}$$
$$20932$$
$$\underline{19200}$$
$$1732$$
$$\underline{1680}$$
$$52$$
$$\underline{48}$$
$$4$$

8.
```
        24
   × 2 872
        48
     1¹2
       48
      3
    ¹1 6 2
      4 8
    6 8 9 2 8
  +       4
    6 8,9 3 2
```

9. $16 \times 8 \times 7 = 896$ cu ft

10. $25 \times 21 \times 20 = 10{,}500$ cu in

11. $32 \div 2 = 16$
$16 \times 1 = 16$

12. $64 \div 8 = 8$
$8 \times 5 = 40$

13. $42 \div 6 = 7$
$7 \times 5 = 35$

14. $\frac{4}{5}$

15. $\frac{1}{3}$

16. $8 \times 12 = 96$ in

17. 99

18. MMCDLIII

Final Test

1.
$$20$$
$$4\overline{)80}$$
$$\underline{80}$$
$$0$$

2.
$$7\text{ r.4}$$
$$7\overline{)53}$$
$$\underline{49}$$
$$4$$

3.
$$81$$
$$8\overline{)648}$$
$$\underline{640}$$
$$8$$
$$\underline{8}$$
$$0$$

4.
```
      79 r.1
  5 )396
     350
      46
      45
       1
```

5.
```
       25 6/25
  25 )631
      500
      131
      125
        6
```

6.
```
      25
    ×25
     125
      10
      40
     6¹25
    +  6
     631
```

7.
```
       21 13/16
  16 )349
      320
       29
       16
       13
```

8.
```
      16
    ×21
      16
      32
     336
    +13
     349
```

9.
```
     5,076 2/6
  6 )30458
     30000
       458
       420
        38
        36
         2
```

10.
```
        6
    ×5 076
        36
        42
     3 0
     30456
    +    2
     30,458
```

11.
```
      686 23/84
  84 )57647
      50400
       7247
       6720
        527
        504
         23
```

12.
```
         84
    ×   686
       ¹4 2
      ¹6 3 8 4
     ¹4 2 4 2
        8 4
      5 7 6 2 4
    +     2 3
      5 7,6 4 7
```

13. $3+7=10$
$10 \div 2 = 5$
$5 \times 5 = 25$ sq ft

14. $4 \times 9 = 36$
$36 \div 2 = 18$ sq ft

15. $10 \times 25 = 250$ sq in

16. $33 \times 15 = 495$ sq ft

17. $45 \times 29 \times 12 = 15,660$ cu ft

18. $27 \div 3 = 9$

19. $40 \div 2 = 20$

20. $20 \div 4 = 5$

21. $5 \times 4 = 20$

22. $4 \times 16 = 64$

23. $1 \times 5,280 = 5,280$

24. $5 \times 2,000 = 10,000$

25. $36 \div 12 = 3$

26. $40 \times 12 = 480$

27. 50

28. 4,000

29. 500
30. $12 \div 3 = 4$
    $4 \times 1 = 4$
31. $21 \div 7 = 3$
    $3 \times 3 = 9$
32. $32 \div 8 = 4$
    $4 \times 5 = 20$
33. $\dfrac{3}{5}$
34. $\dfrac{2}{3}$
35. 2,543,900,000
36. $5 + 12 + 13 + 21 + 24 = 75$
    $75 \div 5 = 15$
37. 2,158
38. MCMLXXV

Word Problems Lesson 15

1. area of garden = $11 \times 13 = 143$ sq ft

   $10 \times 10 = 100$ sq ft needed for one packet

   $2 \times 100 = 200$ sq ft needed for 2 packets

   $200 > 143$; the garden does not have enough space.

2. Use a drawing to show the girls' travels. The distances don't have to be exact.

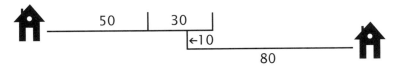

   $50 + 30 = 80$ to turn around

   $80 + 10 = 90$ backtrack to restaurant

   $90 + 80 = 170$ miles = total distance driven

3. $\$50 + \$20 = \$70$ what they left with

   plus $10 to <u>each</u>

   $\$8 + \$15.65 + \$10 = \$33.65$ what they spent

   (gifts = $5+$5)

   $\$70 - \$33.65 = \$36.35$ left

Word Problems Lesson 21

1. $3 \times 75 = 225$ pieces of candy

   $225 \div 7 = 32$ r. 1

   Scott will have one piece left.

2. $5 \times \$3 = \$15$ for babysitting

   $3 \times \$4 = \$12$ for garden work

   $\$15 + \$12 = \$27$

   $\$35 - \$27 = \$8$ more needed to buy the game

3. $25 \div 5 = 5$ acorns in a group

   $16 \div 4 = 4$ seed pods in a group

   $8 \div 2 = 4$ feathers in a group

   $5 + 4 + 4 = 13$ items given to Mom

Word Problems Lesson 27

1. $65 \div 13 = 5$ bags per student

   $5 \times 15 = 75$ nuts per student

2. (You may want to draw this one.)

   biggest move is 5; two times biggest move is 10

   $3 + 5 + 1 = 9$

   $9 - 6 = 3$

   $3 + 10 = 13$; he is 13 spaces from the beginning.

3. $3 \times 11 = 33$

   $V = 33 \times$ missing measure

   $V = 330$ sq in

   $330 \div 33 = 10$, which is the missing measure of the side of the box top.

   $10 \times 11 = 110$ sq in, which is the area of the top of the box.

4. $\$360 \div 6 = \$60$ for each child

   $\$60 + \$20 = \$80 =$ Kate's money

   $\$80 \div 16 = \$5$ cost of each gift

Symbols and Tables

MONEY

1 nickel = 5 cents (5¢)

1 dime = 10 cents (10¢)

1 quarter = 25 cents (25¢)

1 dollar = 100 cents (100¢ or $1.00)

1 dollar = 4 quarters

MEASUREMENT

1 tablespoon (Tbsp) = 3 teaspoons (tsp)

1 quart (qt) = 2 pints (pt)

1 gallon (gal) = 8 pints

1 gallon = 4 quarts

1 liter (L) = 1,000 milliliters (ml or mL)

1 foot (ft) = 12 inches (in)

1 yard (yd) = 3 feet

1 mile (mi) = 5,280 feet

1 meter (m) = 100 centimeters (cm)

1 kilometer (km) = 1,000 meters

1 pound (lb) = 16 ounces (oz)

1 ton = 2,000 pounds

1 kilogram (kg) = 1,000 grams (g)

60 seconds = 1 minute

60 minutes = 1 hour

7 days = 1 week

365 days = 1 year

52 weeks = 1 year

12 months = 1 year

1 dozen = 12

1 cubic foot of water is about 7 gallons.

1 gallon of water weighs about 8 pounds.

PLACE-VALUE NOTATION

$31,452 = 30,000 + 1,000 + 400 + 50 + 2$

EXPANDED NOTATION

$1,452 = 1 \times 1,000 + 4 \times 100 + 5 \times 10 + 2 \times 1$

SYMBOLS

=	equals
≈	approximately equal to
+	plus
–	minus
×	times
·	times
( )( )	times
¢	cents
$	dollars
'	foot
"	inch
<	less than
>	greater than
‖	parallel
∟	right angle
⊥	perpendicular
↔	line
→	ray
—	line segment

$4 \div 2$ 4 divided by 2

$2\overline{)4}$ 4 divided by 2

$\frac{4}{2}$ 4 divided by 2

AREA AND VOLUME

rectangle $A = bh$ (base times height)

parallelogram $A = bh$

triangle $A = \dfrac{bh}{2}$

trapezoid $A = \dfrac{b_1 + b_2}{2} \times h$ (average length of bases times the height)

rectangular solid $V = Bh$ (area of base times height)

                 or $V = l \times w \times h$ (length times width times height)

LABELS FOR PARTS OF PROBLEMS

Addition

```
  25    addend
+ 16    addend
  41    sum
```

Multiplication

```
  33    multiplicand (factor)
×  5    multiplier (factor)
 165    product
```

Subtraction

```
  45    minuend
- 22    subtrahand
  23    difference
```

Division

```
       2    quotient
divisor 2|4  dividend
```

222

Glossary

A–E

acute angle - an angle with a measure greater than 0° and less than 90°

acute triangle - a triangle in which all three angles are acute angles

angle - a geometric figure formed by two rays joined at their origins

area - the measure of the space covered by a plane shape, expressed in square units

Associative Property of Addition - a property that states that the way terms are grouped in an addition expression does not affect the result

average - a measure of center in a set of numbers, usually referring to the *mean*

base - a particular side or face of a geometric figure used to calculate area or volume

Commutative Property - a property that states that the order in which numbers are added or multiplied does not affect the result

congruent - having exactly the same size and shape

cube - a solid with six congruent faces that meet at right angles

degree - a unit of measure for angles; 1/360 of a circle

denominator - the bottom number in a fraction, which shows the number of parts in the whole

dimension - a measurement in a particular direction (length, width, height, depth)

dividend - the number being divided

divisor - a number that is being divided into another

equation - a mathematical statement that uses an equal sign to show that two expressions have the same value

estimate - a close approximation of an actual value

even number - any number that can be evenly divided by two

expanded notation - a way of writing numbers by showing each digit multiplied by its place value

F–M

factor - (n) a whole number that multiplies with another to form a product; (v) to find the factors of a given product

fraction - a number indicating part of a whole

height - the perpendicular distance from the base to the top of a figure

hexagon - a polygon with six sides

inequality - a mathematical statement showing that two expressions have different values

line - a set of connected points that extends infinitely in two directions

line segment - a section of a line bounded by two endpoints

mean - a measure of center found by dividing the sum of a set of values by the number of values

N-P

numerator - the top number in a fraction, which shows the number of parts being considered

obtuse angle - an angle with a measure greater than 90 degrees and less than 180 degrees

obtuse triangle - a triangle in which one of the angles is greater than 90 degrees

octagon - a polygon with eight sides

odd number - any number that cannot be divided evenly by two

origin - another name for the endpoint of a ray

parallel lines - lines in the same plane that do not intersect

parallelogram - a quadrilateral with opposite sides that are parallel and congruent and opposite angles that are congruent

partial product - the result obtained when a number is multiplied by one part of a multiplier; partial products are added to obtain the final product

pentagon - a polygon with five sides

perimeter - the distance around a polygon

perpendicular lines - lines that form right angles when they intersect

place value - the position of a digit which indicates its assigned value

place-value notation - a way of writing numbers that shows the place value of each digit

plane - a flat, two-dimensional surface that extends infinitely in all directions

point - a defined position in space that has no dimensions; represented with a dot

polygon - a closed plane shape having three or more straight sides that do not cross

product - the result when numbers are multiplied

Q-S

quadrilateral - a polygon with four sides

quotient - the result when numbers are divided

ray - a geometric figure that starts at a definite point (called the origin) and extends infinitely in one direction

rectangle - a quadrilateral with two pairs of opposite parallel sides and four right angles

rectangular solid - a three-dimensional shape with six rectangular faces

regrouping - composing or decomposing groups of ten when adding or subtracting

right angle - an angle measuring 90 degrees

right triangle - a triangle with one right angle

Roman numerals - a system used by the ancient Romans in which letters represent numbers

rounding - replacing a number with another that has approximately the same value but is easier to use

skip counting - counting forward or backward by multiples of a number other than one

square - a quadrilateral in which the four sides are perpendicular and congruent

symmetry - having congruent parts facing each other across an axis, with one the reverse of the other

T-Z

trapezoid - a four-sided polygon with a set of parallel sides

triangle - a polygon with three straight sides

unit - the place in a place-value system representing numbers less than the base

unknown - a specific quantity that has not yet been determined, usually represented by a letter

volume - the number of cubic units that can be contained in a solid

Master Index for General Math

This index lists the levels at which main topics are presented in the instruction manuals for *Primer* through *Zeta*. For more detail, see the description of each level at mathusee.com. (Many of these topics are also reviewed in subsequent student books.)

Delta Index